Classroom Companion: Economics

The Classroom Companion series in Economics includes undergraduate and graduate textbooks alike. It welcomes fundamental textbooks aimed at introducing students to the core concepts, empirical methods, theories and tools of the field, as well as advanced textbooks written for students at the Master and PhD level seeking a deeper understanding of economic theory, mathematical tools and quantitative methods.

Kjetil Bjorvatn

Workbook for Microeconomics Made Simple

Exercises and Solutions

 Springer

Kjetil Bjorvatn
Economics
NHH Norwegian School of Economics
Bergen, Norway

ISSN 2662-2882   ISSN 2662-2890 (electronic)
Classroom Companion: Economics
ISBN 978-3-032-06356-4   ISBN 978-3-032-06357-1 (eBook)
https://doi.org/10.1007/978-3-032-06357-1

Translation based on the Norwegian language edition: "Mikroøkonomi: en abc på 1-2-3. Oppgaver, tips og løsningsforslag" by Kjetil Bjorvatn, © Vigmostad & Bjørke AS 2025. Published by Fagbokforlaget. All Rights Reserved.

This Springer imprint is published by the registered company Springer Nature Switzerland AG
The registered company address is: Gewerbestrasse 11, 6330 Cham, Switzerland

If disposing of this product, please recycle the paper.

Preface

Working with exercises is a key part of learning. It challenges you not just to read the theory, but to apply it, and this learning by doing will undoubtedly give you a deeper understanding of microeconomics. I've created 65 exercises for you—five for each chapter—complete with tips and detailed solution guides.

The exercises are closely integrated with the textbook and offer both interesting applications and theoretical challenges. I encourage you to do the exercises more than once—first when you read the relevant chapter, and again as part of your exam preparation. The first time, you might struggle a bit, even after looking at the tips or reading through the solution. But the second time, I'm confident it will feel much easier. You'll recognise which theory is relevant and what the question is really asking for. Along with the material in the textbook, this will give you a solid understanding of microeconomics and how theory can be used to analyse important issues.

The solutions are quite detailed, although I've occasionally skipped a few intermediate steps and often refer to math boxes in the textbook to save space. Still, I recommend working through the calculations step by step rather than copying any recipe from the textbook. Remember: whether in an exam or at work, you'll need to convince your audience that you're in control—and referring to a math box won't do the trick. You are the expert. You're there to show that you *understand* the theory, and know how to apply it, not just that you memorised it.

As in the textbook, the wonderful illustrations were drawn by my father, Bjarne. They brighten up the pages and create a friendly atmosphere, which I believe is good for learning. The drawings also emphasise the connection between this workbook and the textbook, showing that together they make for a learning experience a bit out of the ordinary.

Bergen, Norway
December 2025

Kjetil Bjorvatn

Competing Interests The author has no competing interests to declare that are relevant to the content of this manuscript.

Contents

Income, Prices and Preferences

Anna is working on microeconomics exercises and looking at housing ads

This chapter is about how we as consumers make our choices. The exercises will help you become more familiar with the key tools of consumer theory: the budget line and the indifference curve, and how they interact.

You'll encounter budget constraints and utility functions presented with a slightly different twist than in the textbook, challenging you to think outside the box—and outside the math box too!

© The Author(s), under exclusive license to Springer Nature Switzerland AG 2026
K. Bjorvatn, *Workbook for Microeconomics Made Simple*, Classroom Companion:
Economics, https://doi.org/10.1007/978-3-032-06357-1_1

1.1   Higher Electricity Prices and Shifts in the Budget Line

Anna has an income $I = 1$, which she wishes to spend on food A and living in a well-heated home, her base B. The price of food and of a warm home is initially given by $p_B^l = p_A = 1$, where superscript "l" stands for *low*. She has decided to spend an equal amount of money on housing and food, so that $p_A A = p_B B$.

a. Draw Anna's budget line in a diagram and mark her chosen consumption bundle.
b. Now assume that the price of electricity increases to $p_B^h = 2$, where superscript "h" stands for *high*. What is the slope of the new budget line, and what is Anna's new consumption choice?
c. The government introduces an electricity subsidy that makes the price consumers pay return to its original level. The per-unit subsidy is therefore $s_B = p_B^h - p_B^l = 1$, that is, the difference between the new high price and the previous low price. What is the total cost of this subsidy, S_B, where $S_B = s_B B$?
d. As an alternative to subsidising prices, the government could offer a cash transfer to consumers. Show how giving Anna a cash transfer S (equal in value to the subsidy she would have received, i.e., $S = S_B$) affects her budget line and her consumption choice.

1.2   Anna and Her Friend Deal with Higher Electricity Prices

In the previous exercise, we assumed that Anna wanted to spend the same amount of money on each good. Now we'll take a slightly more formal approach. Anna and her friend Bella are sitting in a café, chatting about tastes and preferences. Both have preferences represented by a Cobb-Douglas utility function $U = A^\alpha B^{1-\alpha}$, where A is food and B is a warm home. Anna spends equal amounts on A and B, while her friend spends three times as much on housing as on food.

a. What is Anna's utility weight α? And what about Bella's?
b. The prices of the two goods are initially $p_B^l = p_A = 1$, and each of them has an income of $I = 1$. What are their consumption levels of A and B? Calculate and illustrate.
c. What would Anna's marginal rate of substitution (MRS) be if she were to choose the same consumption combination as Bella? What would her utility be in this case (round up your solution to two decimal places)? Compare to her optimal solution and discuss.

d. The price of electricity then increases to $p_B^h = 2$. "How much have you turned down the heating in your flat?" Anna asks. "Are you spending less on food now?" Bella wonders. What are the answers to these questions, based on their utility-maximising choices?

1.3   Price Support or Cash Transfer?

There is a debate on how to best support the population in a time of high electricity prices. The government's proposal is to offer a price subsidy, so that consumers face the same price as before. You are a newly hired economist at the Ministry of Finance, and decide to use Anna as an example.

Anna has preferences over food A and a warm home B given by the utility function $U = A^{0.5}B^{0.5}$. Her income is $I = 1$. Before the price increase, electricity cost $p_B^l = 1$, but the price has now risen to $p_B^h = 2$. The price of food remains constant at $p_A = 1$.

You argue that it would be better for Anna to receive this support as cash transfer rather than as price support. Explain why.

1.4   Peanuts + Beer = True Love

Most people would probably agree that peanuts and beer go well together. But the preferred combination may vary from person to person. Assume that preferences over peanuts A and beer B are described by the utility function $U = \min\left(\frac{A}{\alpha}, \frac{B}{1-\alpha}\right)$. The prices of the two goods are p_A and p_B, and income is given by I.

a. Find the optimal consumption of the two goods as a function of the prices, income and the utility weight α.
b. Brian (whom we'll get to know better in the next chapter) prefers beer, with $\alpha = 1/3$, while Anna prefers peanuts, with $\alpha = 2/3$. Assume each of them has an income of $I = 1$, and that prices are $p_A = p_B = 1$. Show Brian's and Anna's choices in a diagram.
c. Now suppose that the price of beer increases to $p_B^h = 2$. Calculate and illustrate how this affects Brian's and Anna's beer consumption, both in terms of quantity and as a percentage of their original consumption.

1.5   How Does an Interest Rate Increase Affect the Demand for Housing and Other Goods?

You have newly graduated from the School of Economics and started working in the analysis department at the Central Bank. One of your tasks is to monitor how demand develops across different markets. The Central Bank has recently raised its policy interest rate, and other banks are expected to follow suit.

Your supervisor is interested in how a change in the interest rate affects demand for housing and other goods and asks you to use microeconomic theory to shed light on the question.

You begin by assuming that an increase in the interest rate will lead to a higher rent on housing, p_B. Moreover, you choose to use a numerical example to illustrate effects, and let household income be given by $I = 1$, which can be spent on a home base B and activities A. Initially, prices are $p_B^l = p_A = 1$, but the interest rate increase causes the price of B to double to $p_B^h = 2$ (where superscript "l" stands for *low* and "h" for *high*).

Housing is a special good—people need a roof over their heads. You therefore choose to study two different utility functions, a balanced Cobb-Douglas utility function $U = A^{0.5}B^{0.5}$ and a utility function where housing is a basic good, $U = A^{0.5}(B - b)^{0.5}$, where b is the minimum required consumption of housing, and where we assume that $b = 0.2$.

Calculate the consumption of A and B before and after the interest rate increase and illustrate your findings using one figure for each utility function. Comment on the differences you observe.

1.6 Tips

1.1 Higher Electricity Prices and Changes in the Budget Line
This exercise helps you become more familiar with the budget line, both what determines its slope and what determines its position. Here, we're interested in an increase in the price of electricity, and two different policy responses: a price subsidy or as a cash transfer.

Remember that a price subsidy affects the slope of the budget line, while a cash transfer leads to a parallel shift of the budget line. When drawing the budget line, it can be useful to identify the maximum consumption of each good, which defines the points where the budget line crosses the axes. You will see that the price subsidy shifts the line back towards its original position before the price increase, while a cash subsidy shifts the budget line outward but maintains the steep slope determined by the higher electricity price.

1.2 Anna and Her Friend Deal with Higher Electricity Prices
This exercise provides practice in understanding the choices of a utility-maximising consumer. Again, it involves different forms of subsidies during a period of high electricity prices and builds on Exercise 1.1, but here we explicitly incorporate the consumer's utility function. It may be helpful to refer to Math Box 1.2 to solve this exercise.

1.3 Price Support or Cash Transfer?
Using the utility function introduced in Exercise 1.2, show the consumer choices under these different subsidy schemes. Then plug the chosen levels of A and B into

the utility function to determine which policy yields the highest utility. Try also to understand your results intuitively.

1.4 Peanuts + Beer = True Love

Some goods naturally go together, such as beer and peanuts, and this exercise gives you practice in analysing situations with perfect complementarity between two goods, that is, Leontief preferences. With this type of preference, utility is determined by the minimum of the two components in the utility function.

Accordingly, the consumer will choose $\frac{A}{\alpha} = \frac{B}{1-\alpha}$ since no other combination yields higher utility.

To solve part (a), use this optimal consumption combination in the budget constraint $I = p_A A + p_B B$ and solve for optimal consumption of the two goods as a function of prices, income and the utility weight α.

Brian and Anna have different views about the ideal mix between the two goods. In part (b), use your result from part (a), applying the relevant utility weights: $\alpha = 1/3$ for Brian and $\alpha = 2/3$ for Anna.

In part (c), consider a situation where the price of good B increases. The condition $\frac{A}{\alpha} = \frac{B}{1-\alpha}$ must still hold, but when you substitute this into the budget constraint with the higher p_B, you'll find a new level of consumption for both goods.

The change in consumption of good B (illustrated here using Brian) is $B_B^l - B_B^h$, when going from the low to the high price. The percentage change is $\frac{B_B^l - B_B^h}{B_B^l}$. The same applies for Anna.

1.5 How Does an Interest Rate Increase Affect the Demand for Housing and Other Goods?

This exercise introduces us to a utility function with a basic good and shows that it leads to some interesting differences in consumption patterns compared to a standard utility function.

You can apply what you've learned about optimal consumption under a balanced Cobb-Douglas utility function, $U = A^{0.5}B^{0.5}$ (see Math Box 1.2), but with $B - b$ substituted for B. For instance, while the marginal rate of substitution based on a balanced Cobb-Douglas utility function is $MRS = A/B$, when B is a basic good it's $MRS = A/(B-b)$.

More About Consumer Choice

Brian enjoys an IPA (whenever he can afford one)

This chapter takes a closer look at how changes in prices and income affect consumer choices, and the exercises here will give you a deeper understanding of the key mechanisms involved.

Depending on the consumer's preferences, a change in price or income may cause demand to increase or decrease, by a little or a lot. Based on this response, we can classify goods as normal, inferior, or Giffen.

© The Author(s), under exclusive license to Springer Nature Switzerland AG 2026   7
K. Bjorvatn, *Workbook for Microeconomics Made Simple*, Classroom Companion:
Economics, https://doi.org/10.1007/978-3-032-06357-1_2

2.1   The Effect of Higher Electricity Prices, in Two Steps

Anna and her friend Bella are still sitting at the café, discussing the high electricity prices. The two goods—food (A) and a warm home (B)—initially cost $p_B^l = p_A = 1$, but the price of electricity then doubles to $p_B^h = 2$. They both have an income of $I = 1$.

Anna has a balanced Cobb-Douglas utility function, while Bella places greater emphasis on having a warm home, with the utility function $U_{Bella} = A^{0.25}B^{0.75}$. Anna's substitution and income effects from the price increase are shown in Math Box 2.1, but what about Bella? Focus on the effect on consumption of good B and compare the impact of the price increase for the two friends using calculus and illustrations.

2.2   Cash Transfers for Education in Mexico: Part I

Many countries have introduced cash transfers for poor families. Sometimes these transfers come with conditions, such as requiring parents to send their children to school. The most well-known example of such a conditional cash transfer pro-gramme was launched in Mexico in the 1990s under the name Progresa. In more recent years, many countries have instead opted for unconditional cash transfers, as this gives families greater freedom to allocate the funds as they see fit.

Consider a family in Mexico—the Buenos family—with preferences over chil-dren's schooling (good B) and other goods (good A), represented by the utility function $U = A^\alpha B^{1-\alpha}$, where $\alpha = 0.5$ (i.e., a balanced Cobb-Douglas utility func-tion). The family's income is $I = 1$, and the prices of both goods are $p_A = p_B = 1$. You can interpret good B as the number of children the family sends to school, where p_B is the school cost per child.

a. What is the family's optimal choice of A and B without any cash transfer? Calculate and illustrate.
b. The government now introduces an unconditional cash transfer of $S = 0.5$, so that the family's income becomes $I' = I + S = 1.5$. How does this affect the family's consumption choice? Calculate and illustrate.
c. Some propose that the transfer should instead be conditional: to qualify for the support $S = 0.5$, the family must send all school-aged children to school and pay the required school fees. Think of this as a minimum requirement: $B_{min} = 0.75$.

What will the choices of the Buenos family be under such a conditional cash transfer? Calculate, illustrate, and comment on your findings in comparison with the unconditional transfer in part (b).

2.3   Cash Transfers for Education in Mexico: Part II

Now consider the Aires family, who has the same income as the Buenos family but place less weight on children's education in their utility function: $U = A^{\alpha}B^{1-\alpha}$, where $\alpha > 0.5$. Otherwise, the situation is exactly as for the Buenos family, with income $I = 1$, and prices $p_A = p_B = 1$.

a. How will the Aires family's choices be affected by an unconditional and a conditional cash transfer, with the same terms as those discussed for the Buenos family? Illustrate using diagrams (no calculations needed), and comment on your findings. Will the Aires family always choose to accept the conditional cash transfer?

b. There is ongoing debate over which is better: conditional or unconditional support. Three common claims in this debate are:
   A. It is necessary to make the transfer conditional to achieve the government's goals for schooling.
   B. Making the support conditional on schooling makes the cash transfer less valuable to families.
   C. It makes no difference whether the support is conditional or unconditional.
   What can you say about these claims considering your analysis of the Buenos and Aires families?

2.4   Brian's Choice: Beer or Wine... or Both?

Brian enjoys both beer (B) and wine (A). His preferences are given by the utility function $U = A + lnB$. This is a type of utility function we haven't worked with before (it's called quasilinear), so feel free to refer to the tip if you need help getting started.

   We shall in this exercise focus on how a change in income affects Brian's choice of beer and wine and therefore assume that $p_B = p_B = 1$, so that the budget constraint is given by $I = A + B$.

a. What is Brian's marginal utility of wine, and what is his marginal utility of beer?
b. At what income level will Brian start buying wine? Use a diagram (or two) to show how the consumption of beer and wine changes as income increases, that is, the income-expansion line. (Hint: place income on the horizontal axis, and consumption of beer and wine on the vertical axis.)
c. What if Brian's preferences for beer and wine were instead described by a balanced Cobb-Douglas utility function, $U = A^{0.5}B^{0.5}$? How would the relationship between income and consumption look in that case? Compare with what you found under the quasilinear utility function and draw the income-expansion line in the figure.

d. Would you say that beer is an inferior good for Brian, given quasilinear preferences?

2.5 Conrad's and Anna's Saving

Conrad and Anna are thinking about how much to save. For Conrad, it's saving for retirement, while for Anna, it's about saving for the next semester. Assume that they both have a balanced Cobb-Douglas utility function $U = x_2^{0.5} x_1^{0.5}$, where x_1 and x_2 are consumption in period 1 and 2.

For Conrad, period 1 is his last working years, which gives him an income $I_1^C = 1$, while his reteriment is period 2, with zero income: $I_2^C = 0$. This implies that whatever he consumes during his retirement must come from savings from current income.

Anna, on the other hand, works besides here studies and has income in both periods $I_1^A = I_2^A = 0.5$.

a. Initially, the interest rate is zero, $r = 0$. What are Conrad's and Anna's savings? Calculate and illustrate, with one figure for Conrad and one for Anna.
b. Now assume that the interest rate increases to $r' = 0.25$. What happens to their saving decisions?
c. Compare the effect of an interest rate increase on their saving and explain the difference.

2.6 Tips

2.1 The Effect of Higher Electricity Prices, in Two Steps
Math Box 2.1 is useful for solving this problem. It shows the procedure for finding the substitution and income effects for a balanced Cobb-Douglas function, while here you are asked to find it for the general Cobb-Douglas utility function. You'll need to stay focused, but with a few tips, you'll be fine!

The first tip is to look at Exercise 1.3, which has the same starting point and asked you to look at the same price change (but without decomposing the price effect). With $p_B = p_A = 1$ and income $I = 1$, we know that the optimal consumption is $A = \alpha$, $B = 1 - \alpha$. We can insert these values into the utility function and find the initial utility level as:

$$U = A^\alpha B^{1-\alpha} = \alpha^\alpha (1 - \alpha)^{1-\alpha}$$

Beyond that, you should follow the three steps for the substitution and income effects, as described in the textbook. You can continually check your calculations by setting $\alpha = 0.5$; if you get the same result as in Math Box 2.1, you're on the right track!

2.2 Cash Transfers for Education in Mexico: Part I

This exercise challenges you to think about how different types of support schemes affect the budget line. Unconditional support should be straightforward: it simply results in a parallel outward shift of the budget line in the diagram.

With *conditional* support, the budget line will have a kink: there is no support as long as $B < B_{min}$, and then a positive support (a parallel shift of the budget line) for $B \geq B_{min}$.

Note that the Buenos family places equal weight to consumption of other goods and schooling in their utility function. Does conditionality matter in this case?

2.3 Cash Transfers for Education in Mexico: Part II

The Aires family derives more uility from other goods, and therefore spends less on their children's schooling before the support. And, as you will see, this family's optimal choice is affected by whether the support is unconditional or conditional.

Will the Aires family necessarily want to accept the conditional cash transfer? One can imagine that if Aires places sufficiently high value on other goods relative to their children's education, they would be better off without the grant than with it, since the conditions force them into a completely different consumption pattern than what they ideally want. In the diagram, you can show this by an indifference curve that is tangent to the budget line without support and which also passes through the corner solution with support, where the family spends $B = B_{min}$ and the rest of the money on A. From the tip to Exercise 2.1 above, we know that the utility without the grant is given by:

$$U = A^{\alpha} B^{1-\alpha} = \alpha^{\alpha} (1 - \alpha)^{1-\alpha}$$

Compare this with utility with the grant, and you can find the critical α (Hint: The number is quite close to one: almost all weight must be placed on A for Aires to say no to the grant!).

2.4 Brian's Choice: Beer or Wine... or Both?

In part (a), you are asked to find the marginal utility of each of the two goods, and here you will need the differentiation rule:

$$\frac{\partial lnX}{\partial X} = \frac{1}{X}$$

Evidently, the marginal utility of beer is much higher than the marginal utility of wine for small quantities of beer (when B is low). This means that Brian will start with beer before moving on to wine.

When will he start thinking about wine? That happens when the marginal utility of wine is equal to the marginal utility of beer. So, use the condition $MU_B = MU_A$ to find the critical income level at which Brian starts buying wine.

At this point, he gets just as much satisfaction from one more sip of beer as from his first sip of wine—we can say that he has reached his "beer target".

We also know that for any quantity of beer less than this, he will spend all his money on beer. With this information, you can find the critical income level required for Brian to start buying wine.

On part (c), consumption guided by Cobb-Douglas preferences is something you're well familiar with, but check Math Box 1.2 if you're unsure.

As for inferior goods, remember that these are goods that people buy less of when their income rises. Does that apply to beer in this case?

2.5 Conrad's and Anna's Saving

On part (a), note that Conrad has income only in period 1, so for him, an interest rate increase has the same effect as a decrease in the price of consumption in period 2: the budget line rotates around the point where it intersects the horizontal axis.

Anna has the same income in both periods, and since she places equal weight on consumption in both periods and the interest rate is zero, it's intuitive that saving is zero—there's no reason to shift consumption in one direction or the other.

Part (b) deals with an increase in the interest rate. For Conrad, this effect works like a reduction in the price of x_2. So, the impact on his consumption in period 1, and hence his savings, is a study of the cross-price effect. And with Cobb-Douglas preferences, we know what happens (nothing changes with consumption in period 1!).

For Anna, an increase in the interest rate causes her budget line to pivot around the point of zero saving. It's like a combination of a lower price for consumption in period 2 (the intercept on the vertical axis moves up) and a higher price for consumption in period 1 (the intercept on the horizontal axis moves left).

The sum of these two effects on saving can be found using the expression for optimal saving in Math Box 2.2. You'll see that it leads to lower consumption in period 1, in other words, more saving. So now it's just a matter of plugging in the given numbers and drawing a nice figure!

In part (c), you are asked to contrast the impact on saving for our two friends. The key to understanding this difference lies in understanding how the interest rate increase affects their budget lines.

Consumers at Work **3**

How much time should Anna Spend working?

This chapter deals with labour supply, which is an important application of consumer theory. Economists view labour supply as a choice between consumption and leisure. While consumption is limited by the budget, leisure is limited by time: there are only 24 h in a day and seven days in a week, which sets the stage for the choice between leisure and working hours. The exercises in this chapter will help you become more familiar with both the decision to work and the choice of how much to work under different assumptions about preferences, wages, and non-labour income.

© The Author(s), under exclusive license to Springer Nature Switzerland AG 2026   13
K. Bjorvatn, *Workbook for Microeconomics Made Simple*, Classroom Companion:
Economics, https://doi.org/10.1007/978-3-032-06357-1_3

3.1   Audrey Enjoys Working, While Beth Works Just for the Money

Audrey and Beth are colleagues at the office. They have a lot of flexibility in choosing work hours and are thinking about how to allocate their time, between work and leisure. They both have a time endowment of $T = 18$, the same wage $w = 1$ and face the same price $p = 1$. Let M be consumption, F leisure, and J hours worked.

Beth works only for the money, and has a standard, balanced Cobb-Douglas utility function:

$$U_B = M^{0.5}F^{0.5} \quad \text{Beth's utility function}$$

Audrey, on the other hand, has worked at the same place for a long time and enjoys her job, as reflected in her utility function:

$$U_A = M^{\frac{1}{3}}F^{\frac{1}{3}}J^{\frac{1}{3}} \quad \text{Audrey's utility function}$$

Working hours (J) enter positively into Audrey's utility—she enjoys working! Note that Audrey's utility function can alternatively be written as (which is more compact and easier to work with):

$$U_A = (MFJ)^{\frac{1}{3}} \quad \text{Audrey's utility function}$$

How much will Beth and Audrey choose to work? Calculate and illustrate using figures.

3.2   Reservation Wage

You work as a freelancer for a financial firm, and the job gives you complete flexibility in choosing your work hours. Your utility function is given by $U = M^{0.5}F^{0.5}$, the total available time per day is $T = 16$, the price of consumption goods is $p = 1$, and your hourly wage is $w_a = 2.25$.

a. How many hours per day do you want to work, and how much consumption and utility will this give you?
b. The manager has offered you a leadership position in the firm, with responsibilities that would require you to work 12 h per day. What is your reservation wage for accepting this job? Calculate, illustrate in a diagram, and discuss your findings.

3.3   Labour Supply with Cash Transfers

Anna derives utility from consumption M and leisure F, represented by the balanced Cobb-Douglas utility function $U = M^{0.5}F^{0.5}$. She receives a regular cash gift S from her grandfather, so her budget constraint is given by:

$$w(T - F) + S = pM$$

Assume $p = 1$ and $T = 16$, so the budget constraint can be written as $w(16 - F) + S = M$.

a. Derive an expression for Anna's labour supply as a function of the wage w and the cash transfer S.
b. What is her labour supply when $w = 1$ and $S = 0$? What if $S = 6$? Is there a level of S that would make her not want to work at all? Calculate and illustrate.

3.4   The Shop Manager Thinks Anna is Responding Strangely to a Wage Increase

The manager at the grocery shop has twice offered Anna a higher wage to encourage her to work more hours during busy weeks. His offer is a 50% wage increase, from the standard $w = 1$ to $w = 1.5$, but is surprised by her response. As in Exercise 3.3, Anna's utility function is $U = M^{0.5}F^{0.5}$, $p = 1$ and $T = 16$.

a. The first time Anna got a wage increase, she was not receiving any support from her grandfather ($S = 0$). The manager notices that the higher wage does not make her want to work more, and he cannot understand why. Calculate, illustrate with a diagram, and explain Anna's choice.
b. The second time the manager increased her wage, Anna was receiving a cash transfer of $S = 6$ from her grandfather. Now the manager is completely confused, as he observes that the wage increase makes Anna want to work more. Calculate, illustrate with a diagram, and explain Anna's choice.
c. Why does the wage increase affect Anna's labour supply differently in the two situations (i.e., without and with the cash transfer)? Try to provide an intuitive explanation.

3.5   Universal Basic Income and Labour Supply

The welfare state ensures a minimum level of consumption for its citizens through various benefit schemes for those in need. Some are concerned that this may reduce people's willingness to work. Could there be some truth to this?

Consider two individuals, Mark and Frank, who both have the utility function $U = M^{0.5}F^{0.5}$, and where $T = 16$, $w = p = 1$.

a. What is their labour supply in the absence of welfare benefits?
b. The government introduces a welfare scheme in the form of transfer S which guarantees a minimum consumption level $M_0 = 4$. Show that Mark and Frank are indifferent between working and living solely on welfare in this case. Illustrate with a diagram.

Assume now that Mark chooses to work, while Frank chooses to live on welfare. The government is concerned that welfare may reduce labour supply and is considering introducing a universal basic income instead, a cash transfer $S = 4$, given unconditionally to everyone. (Hint: the budget constraint would then become $w(T - F) + S = pM$).

c. How is Frank's labour supply affected by the shift from the minimum consumption scheme to universal basic income?
d. How is Mark's labour supply affected by the shift from the minimum consumption scheme to universal basic income?
e. Compare your findings for Frank and Mark and comment on the difference.

3.6   Tips

3.1 Audrey Enjoys Working, While Beth Works Just for the Money
In analysing labour supply, we usually assume that people work only to earn money. But many people probably also experience a sense of satisfaction from being able to go to work.

How does this affect the choice of working hours? This exercise shows how we can incorporate job satisfaction with a small adjustment to the standard theory.

Beth's utility function is a standard, balanced Cobb-Douglas utility function, so you should be able to find her labour supply without too much difficulty (see Math Box 3.1).

Audrey's utility function, $U_A = (MFJ)^{\frac{1}{3}}$, may look a little unusual, but you can think of it as a balanced Cobb–Douglas function, only with three arguments instead of two. We insert $J = T - F = 18 - F$ and then the utility function becomes:

$$U_A = (MF(18 - F))^{\frac{1}{3}}$$

Now we again have a utility function with two arguments, consumption M and leisure F, and we use the standard approach to find the labour supply, as shown in Math Box 3.1. We begin by deriving the marginal rate of substitution (MRS), which is given by:

$$MRS = \frac{MU_F}{MU_M}$$

Differentiate with respect to F (and use the chain rule) to find the marginal utility of leisure U_F:

$$\frac{\partial U_A}{\partial F} = \frac{1}{3}(MF(18-F)^{\frac{1}{3}-1}(18M-2MF)$$

And differentiate with respect to M to find the marginal utility of income MU_M:

$$\frac{\partial U_A}{\partial M} = \frac{1}{3}(MF(18-F)^{\frac{1}{3}-1}(F(18-F))$$

In the optimum, $MRS = \frac{w}{p}$, and that should be enough to get you started!

3.2 Reservation Wage

You are considering switching to a management position and feel that you must be at least as well off as you are today if you're going to work more hours.

You currently work 8 h ($J_a = 8$), which with a wage of $w_a = 2.25$ gives you an income (and consumption) of $M_a = w_a J_a = 18$. Your available time is $T = 16$, so your current leisure is $F_a = T - J_a = 8$.

With the utility function $U = M^{0.5}F^{0.5}$, this means that your current utility is $U_a = (18)^{0.5}(8)^{0.5} = 12$.

A management job (b) must match this level of utility. You can find the income M_b that makes you equally happy by solving:

$$U_b = (M_b)^{0.5}(F_b)^{0.5} = U_a = 12$$

We know the management job requires you to work $J_b = 12$, which means $F_b = T - J_b = 4$.

You can now find the income level required to get you to switch jobs, and from that, you can also determine the reservation wage.

3.3 Labour Supply with Cash Support

You'll need Math Box 3.1 for this exercise. With a cash transfer S, we must rewrite the budget constraint, which (adding the transfer) now becomes $w(T-F)+S = pM$. That is, consumption is financed both by labour income $w(T-F)$ and the transfer S.

We are told that $T = 16$ and $p = 1$, so the budget constraint can be written as:

$$w(16-F)+S = M$$

We found in Math Box 3.1 that the optimal consumption combination can be expressed as:

$$M = \frac{F\alpha}{(1-\alpha)}\frac{w}{p}$$

With $\alpha = 0.5$ and $p = 1$, this simplifies to:

$$M = wF$$

We insert this expression into the new budget constraint and get:

$$w(16 - F) + S = wF$$

You now have the starting point for finding Anna's choice of leisure, and from this, you can find working hours as $J = T - F = 16 - F$. When working with this exercise, note how the cash transfer enters the decision about labour supply.

3.4 The Shop Manager Thinks Anna Reacts Strangely to a Higher Wage

This exercise builds directly on the previous one. There, you found the expression for the optimal choice of working hours with cash support. How does a wage increase affect the labour supply in this case?

When you are asked to make a diagram in part (b) that illustrates the effect of the wage increase, you'll end up with four budget lines. The first two are the same as in Exercise 3.3: m_1 (low wage, no support) and m_2 (low wage, with support). But now, you also get the budget lines m_3 (high wage, no support) and m_4 (high wage, with support).

You can identify the choices in the different scenarios on these budget lines.

When asked to explain the intuition in part (c), it can be helpful to think about the fact that she has two sources of income, one from work and one independent of work.

While an increase in non-labour income has a pure income effect, which pulls in the direction of less work, a wage increase also contains a substitution effect, which pulls in the direction of more work.

3.5 Universal Basic Income and Labour Supply

Many people are concerned that welfare benefits make it more attractive to live on welfare than to get a job. Math Box 3.1 is helpful here. In the first part of the exercise, note that welfare makes it possible to consume $M_0 = 4$ without working, so you must compare the utility of the two individuals in the exercise when working versus living on welfare (the minimum consumption scheme). The alternative welfare support, a universal basic income, is a cash transfer, and affects labour supply in the same way as described in Exercises 3.3 and 3.4. This is a good example of how standard microeconomic theory can be used to shed light on an important and politically controversial issue.

Behavioural Economics

Altruistic Audrey sponsors her son

Behavioural economics challenges some of the standard assumptions about how we make choices, for example, the assumption that our preferences are stable over time. In these exercises, you are encouraged to reflect on how bounded rationality, as well as social preferences (such as altruism), can influence our choices, for example, Audrey's choice between spending money on herself and on her son Brian.

© The Author(s), under exclusive license to Springer Nature Switzerland AG 2026
K. Bjorvatn, *Workbook for Microeconomics Made Simple*, Classroom Companion: Economics, https://doi.org/10.1007/978-3-032-06357-1_4

4.1   Patient and Impatient Brian

Assume that Brian has preferences given by $U = x_2^\alpha x_1^{1-\alpha}$, where the patient version of Brian, version A, has $\alpha = 0.5$, while the impatient version B places greater weight on consumption today than in the future, with $\alpha = 0.25$. Assume also that Brian only receives income in period 1, $I_1 = 1$, and that the interest rate is zero, $r = 0$.

a. How much does patient Brian A want to save? Calculate and illustrate.
b. How much does impatient Brian B want to save? Calculate and illustrate in the same diagram.
c. What is the utility loss, from the perspective of patient Brian, if impatient Brian gets his way? Round to the nearest two decimals.

4.2   Brian Ties Himself to the Mast

This exercise continues where the previous one left off. Patient Brian has thought about how to make sure his impatient self will not be tempted to empty the savings account. He has chosen to open a savings account that charges a fixed penalty g if any money is withdrawn from it in period 1.

a. How does such a penalty affect the budget line? Illustrate using a diagram.
b. How would a penalty affect impatient Brian's utility? Use the hint if you're unsure how to go about this.
c. How large must the penalty be to ensure that patient Brian A can be confident that impatient Brian B will not withdraw from the savings account? Calculate and illustrate in the diagram.

4.3   The Taxi Drivers in New York

Allan and Bill drive taxis in New York. They each have a total of $T = 16$ hours available per day, which must be allocated between work and leisure. They earn a wage w per hour, and the price of consumption goods is $p = 1$.

Allan has a utility function given by $U_A = M^{0.5}F^{0.5}$, where M is consumption of material goods and F is leisure. Allan chooses how much to work based on utility maximisation. Bill, on the other hand, determines his labour supply based on an income target of $M_B^* = 8$.

a. Calculate the labour supply of Allan and Bill.
b. Suppose the drivers work six days a week, from Monday to Saturday, and the hourly wage varies depending on how busy the day is. The week can be divided into three scenarios:

Scenario 1. Monday and Tuesday. Few customers, low wage: $w_1 = \frac{2}{3}$

Scenario 2. Wednesday and Thursday. More customers, moderate wage: $w_2 = 1$

Scenario 3. Friday and Saturday. Many customers, high wage: $w_3 = \frac{4}{3}$.

What is the labour supply of the two drivers on the different days? Calculate and illustrate (with one figure for each scenario).

c. Which of the two drivers works the most, and who earns the most over the course of a week?

4.4 Audrey's Altruism

Refer to Fig. 4.6 in the textbook, which analyses Audrey's altruism and how much she chooses to give to her son, Brian, depending on whether he works or not. Assume Audrey has an income of $I_A = 1$, and that her utility function is $U = A^{0.5}B^{0.5}$, where A is Audrey's own consumption and B is Brian's consumption. Let the price of the consumption good be the same for both and equal to $p = 1$.

a. Assume that Brian has no income at all. What is the mother's optimal allocation in this situation? In other words, what is point a in Fig. 4.6 of the textbook?

b. Now assume that Brian earns his own income, $I_B = 0.25$. How does this affect the mother's allocation (how much she spends on Brian and how much she spends on herself), and what is Brian's total consumption?

4.5 Give Smarter! Anna Offers Audrey Some Sound Advice

Anna has had a serious talk with Audrey. She says that the gifts are making Brian lazy (that is, he works less). Instead of a cash transfer, Anna suggests that Audrey should offer Brian a wage subsidy (that is, a gift that is proportional to his earned income). This, Anna says, could allow Brian to reach the same level of consumption, but at a lower cost to Audrey.

Assume that Brian has Cobb-Douglas preferences over leisure and consumption and use what you learned in Chapter 3 about consumers and labour supply to highlight Anna's point. (Note: This is a graphic exercise, no math needed.)

4.6 Tips

4.1 Patient and Impatient Brian

This exercise is based on Chapter 4.2 in the textbook, see in particular Fig. 4.1. Here, we will study this more closely using a numerical example.

You'll need the information from Math Box 2.2 on intertemporal consumption choice. The Math Box uses a general Cobb-Douglas function, but here you should insert the specific information provided in the exercise about Brian's utility weights, as well as his income and interest rate. Except for that, the exercise should be relatively self-explained.

4.2 Brian Ties Himself to the Mast

Cobb-Douglas preferences result in period 1 consumption $x_1 = (1 - \alpha)I_1$ and in period 2, $x_2 = s = \alpha I_1$. Now, with a fee, income becomes $I_1 = 1 - g$. Insert this into the utility function, and you will find utility as a function of the fee. This should help you find the solution to part (b).

To find the critical level of g in part (c), you should set this equal to the utility Brian B would get if he does not touch the money in the savings account—i.e., the amount Brian A had placed there. You already know from Exercise 4.1 that Brian A's savings gives him a balanced consumption over time. How much utility does Brian B get from such a consumption profile? Once you've found that utility level, call it U_B^a, set $U_B(g) = U_B^a$, and solve for g. This is the critical value of the fee that the exercise is asking for. Round your answer to the nearest two decimals.

4.3 The Taxi Drivers in New York

From Math Box 3.1, we know that Allan, who has a balanced Cobb-Douglas utility function, supplies the following amount of labour:

$$J_A = T - F = \frac{1}{2}T$$

Bill, on the other hand, has an income target of $wJ_B = M_B^* = 8$, which gives the following labour supply:

$$J_B = \frac{8}{w}$$

We see that Bill must work a lot when the hourly wage is low to reach his income target.

4.4 Audrey's Altruism

In this exercise, we examine the relationship between Brian and his mother, Audrey. This builds on Exercise 3.5, where we saw how a given gift, in the form of welfare benefits, affected labour supply. But now, the gift is not fixed, it is something Audrey decides based on her altruistic utility function.

The focus here is on the giver, not the receiver. How much will she give, and what happens to the gift if Brian gets a job?

You can use Math Box 1.2 to find Audrey's optimal choice of goods A and B, where B represents her son's consumption (which enters her utility function). In part (b), consider a situation where Brian earns his own income (I_B), meaning he pays for part of his consumption himself. Audrey's budget constraint can then be written as $I = A + (B - I_B)$.

Audrey pays for the portion $(B - I_B)$ of Brian's consumption, while he covers the rest, I_B. With her balanced Cobb-Douglas utility function, Audrey's utility maximising consumption combination is $A = B$. Starting from this, find how much she now wants to give her son.

4.5 Give Smarter! Anna Offers Audrey Some Sound Advice

Gifts can be complicated, and one consequence may be that they reduce people's willingness to earn their own income. Here you will use what you learned about labour supply in Chapter 3 in the textbook. Pay special attention to Fig. 3.4, which shows the effect of a higher wage, and Fig. 3.5, which illustrates the effect of a cash transfer on labour supply and consumption.

Note particularly that with Cobb-Douglas preferences, and with work as the only source of income, labour supply is unaffected by changes in the wage, and therefore also unaffected by a wage subsidy (which works just like a higher wage by making the budget line steeper).

In your figure, show which budget line makes Brian achieve the same consumption with a wage subsidy as with a cash transfer, and compare how much Audrey needs to sponsor her son to achieve this consumption in the two cases.

Labour and Capital

5

Labour and capital at The Mill

This chapter deals with how producers choose a combination of inputs to keep production costs as low as possible for a given output level. It also addresses the choice of technology: the robots are coming, and they may challenge traditional ways of doing things. The exercises will help you become more familiar with the key tools in this analysis, namely the *isocost* and the *isoquant*. How do changes in

© The Author(s), under exclusive license to Springer Nature Switzerland AG 2026
K. Bjorvatn, *Workbook for Microeconomics Made Simple*, Classroom Companion:
Economics, https://doi.org/10.1007/978-3-032-06357-1_5

input prices affect the choice of input mix? Should firms respond to high labour costs by moving production to a low-cost country, or by investing in new, labour-saving technology at home?

5.1   Same Output or Same Cost?

A firm has a production function $Q = K^{0.5}L^{0.5}$ and an output level $Q = 1$. The initial input prices are $w = r = 1$.

a. What input combination will the firm choose, and what will the production cost be?
b. The wage now increases to $w = 2$. Assume the firm wants to keep output constant ($Q = 1$). How does the wage increase affect the firm's input choice? Calculate (round to the nearest decimal) and illustrate.
c. Assume instead that the firm wants to keep costs constant, while allowing Q to vary. How does the wage increase (from $w = 1$ to $w = 2$) affect the firm's input choice in this case? Calculate (round to the nearest decimal) and illustrate.

5.2   How Much Does Conrad Save by Moving Production to China?

How much can be saved by relocating production to China? Assume that the wage in Conrad's home country is $w_H = 1$ and in China $w_C = 0.25$, while the price of capital is the same in both countries, $r = 1$. Conrad has decided to produce $Q = 1$ (that is, one billion sheets of paper), regardless of location. The production technology is given by $Q = K^{0.5}L^{0.5}$.

a. What are the cost savings from moving production to China if Conrad uses the same input combination there as at home? Calculate and illustrate.
b. What is the optimal input combination for production in China? How much additional cost savings can be achieved by adjusting the input mix to the lower wage level in China? Calculate and illustrate in the same figure as in part (a).

5.3   Leontief & Sons

Leontief & Sons is renowned for its high-quality men's shirts. The company was founded by the Russian immigrant Wassily Leontief, who came to the United States in the early 1900s. The production is (perhaps not surprisingly) based on a Leontief production function:

$$Q = min\left(\frac{K}{\alpha}, \frac{L}{1 - \alpha}\right)$$

Production today takes place in three different factories, with Wassily's three sons in charge. The factories were established at different points in time and use slightly different technologies:

$$\alpha_1 = \frac{1}{3} \quad \text{First factory}$$

$$\alpha_2 = \frac{1}{2} \quad \text{Second factory}$$

$$\alpha_3 = \frac{2}{3} \quad \text{Third factory}$$

Today, each factory produces the same quantity of shirts, $Q = 1$, and the Leontief brothers wish to continue doing so at the lowest possible cost. The wage in the USA is $w_U = 1$, and the price of capital is $r_U = 1$. To keep costs down, the brothers are considering relocating one or more of the factories to China. The wage level there is lower, $w_C = 0.5$, but political tensions lead to a risk premium (investors demand a compensation for investing there), such that the cost of capital in China is $r_C = 1.5$.

a. What location advice would you give the Leontief brothers?
b. Now assume the political climate improves, such that the price of capital in China falls to $r_C = 1.25$. How does this affect your relocation advice? Calculate and illustrate.

5.4   On the Robotisation of Paper Production

Conrad has recently attended the yearly paper conference in Stuttgart and is inspired by all the talk about robotisation in the industry. Today, he produces a given amount of paper, $Q = 1$, using traditional technology $Q = K^{0.5}L^{0.5}$.

With robot technology, the production function becomes $Q = AK$, where the current robot efficiency is $A = 0.5$.

a. Assume $w = r = 1$ and that output remains at $Q = 1$. Would you recommend Conrad to switch to robot technology? Calculate and illustrate.
b. Now assume the price of capital increases to $r = 4$. How does this affect your recommendation?
c. Imagine the robots become more efficient, so $A^{new} > 0.5$. How high must A^{new} be for it to be profitable to switch to robot production when $r = 4$?

5.5   Robots or China?

Conrad produces one billion sheets of paper $Q = 1$ using the traditional technology $Q = K^{0.5}L^{0.5}$. Alternatively, he could use robots with $Q = AK$, where robot efficiency is $A = 0.75$. And then of course there's the option of producing in China.

The wage level in his home country is $w_H = 1$, while in China it is $w_C = 0.5$. The price of capital at home is $w_H = 1$, while heightened political uncertainty in China has resulted in a risk premium, so that $r_C = 2$.

You are asked to advice Conrad on both choice of technology and location, where you are told that the quantity produced should stay the same.

a. Which type of technology would you recommend if he stays at home: robot or traditional?
b. Which type of technology would you recommend if he were to move production to China: robot or traditional?
c. Compare your findings from parts (a) and (b): Where would you recommend Conrad to locate production?

5.6   Tips

5.1 Same Output or Same Cost?
This exercise invites you to study how a change in one of the factor prices affects the optimal factor composition and production costs. Usually, we assume the firm's objective is to minimise costs for a given level of output, but here you'll also explore an alternative objective: keeping costs unchanged. Math Box 5.2 gives you the necessary information.

5.2 How Much Does Conrad Save by Moving Production to China?
Conrad is considering relocating production to China, tempted by its low labour costs. There are savings both because the lower wages reduce costs for a given factor mix, and because the low wage level in China makes it attractive to shift production in a more labour-intensive direction. This exercise trains you to understand the implications of changed factor prices for optimal factor composition, and how much cost savings lie in adapting the production method to local conditions. Again, Math Box 5.2 is helpful.

5.3 Leontief & Sons
The goal of this exercise is to give you a better understanding of Leontief production functions, where, as you know, there is perfect complementarity between labour and capital. In this case, the fixed ratio between the two production factors differs across the three factories. The question is how factor prices affect production costs, and again, the application is a relocation decision: China has the advantage of lower

wages, but political risk means that the cost of capital is higher than in the US. Which of the brothers' factories would you recommend relocating to China?

Note that the production function $Q = min\left(\frac{K}{\alpha}, \frac{L}{1-\alpha}\right)$ means that the optimal factor composition must be $\frac{K}{\alpha} = \frac{L}{1-\alpha}$. For $Q = 1$, this implies $K = \alpha$ and $L = 1 - \alpha$.

5.4 On the Robotisation of Paper Production

The robots are coming! But that doesn't necessarily mean it's profitable to automate production. It depends, among other things, on how productive the robots are and on factor prices. This exercise gives you practice comparing costs of producing a given quantity using different technologies and assessing the profitability of adopting new technology.

a. Start by finding the optimal factor combination using traditional technology, calculate the cost, and identify the point where the isocost line crosses the vertical axis (cost measured in capital units). Then show the capital requirement with robot technology (R_1) and choose the technology that gives the lowest cost (measured in capital units).
b. How are production costs affected by an increase in the price of capital, again measured in capital units? Use the same approach as above: compare the point where the new isocost line intersects the vertical axis with the robot requirement R_1, and choose the technology with the lowest intersection point.
c. With the higher price of capital ($r = 4$), what level of A (robot productivity) would make the firm indifferent between the two technologies? Remember that to produce $Q = 1 \equiv Q_1$, the robot requirement is $R_1 = 1/A$. Find the value of A that makes this equal to the vertical axis intercept of the isocost line you found in (b) above.

5.5 Robots or China?

This exercise builds on the previous one about the choice of technology, but now also includes the question of location: Should Conrad keep production at home, where wages are high but the cost of capital is low, or in China, which has lower wages, but where one must pay a risk premium on capital?

The approach is the same as in Exercise 5.4.

Costs

6

Anna, Brian and Conrad discuss expansion

Conrad is, as you know, thinking about expansion, and this chapter will help him understand the costs involved. It is important to distinguish between the short and long run, and the concept of marginal costs is central. The exercises give you practice in deriving and interpreting cost functions under different assumptions about factor costs, production technology, and time perspective. We will visit many companies, a seamstress in Dar es Salaam, the startup Alpha Tech, the Leontief brothers (again), and a cardboard manufacturer considering investing in green technology.

K. Bjorvatn, *Workbook for Microeconomics Made Simple*, Classroom Companion: Economics, https://doi.org/10.1007/978-3-032-06357-1_6

6.1   The Seamstress in Dar es Salaam

You are an intern for the aid organization Oxfam and have been sent to Tanzania to prepare a report on women in business. During this visit, you meet Grace. She lives in Dar es Salaam, and like many women there, she runs a small business. From her kitchen, she sews kanga, a traditional garment for women. Grace owns two sewing machines that she inherited from her mother; one she places on the kitchen table after breakfast, and the other is kept in a cupboard.

During a workday, she manages to produce one garment, with the sewing machine running almost continuously. She does not earn much money, but given how little money she is spending (the kitchen is hers, the sewing machines were inherited, and she does not pay wages to anyone, not even herself), she still thinks she's doing okay. But is she?

You have made a round in the neighbourhood, and it turns out that there's a rental market for sewing machines and other seamstresses are looking for workers for kanga production.

a. As an economist, what costs would you say are relevant in Grace's small business?
b. Grace has been offered to rent some production premises further down the street, and she is considering sewing the clothes there instead. From a cost perspective, does it matter whether she stays in the kitchen or moves production outside the house?
c. What advice would you give to Grace? Should she continue running her own business?

Note: This is an open discussion question and does not require calculations or figures.

6.2   Visiting Alpha Tech

You've been hired by a major consulting firm and as your first assignment, you're asked to visit to the startup Alpha Tech. They want to expend production and need some help in understanding what this would cost. The firm wants to act quickly and they therefore have a short-term perspective (implying that capital is fixed).

You have been given the following background information, which you use when preparing your report: (i) production function $Q = K^{0.5}L^{0.5}$; (ii) installed capital $K_0 = 1$; (iii) the price of labour and capital $w = r = 1$; (iv) the use of intermediate goods is so small that you can neglect it ($z = 0$).

a. Report on the firm's marginal cost and average total cost. Calculate and illustrate in a figure.

One year later, you are again invited to Alpha Tech to talk about costs (they were really happy with your first report!). In the meantime, the firm has invested in more production capital, which has doubled so that $K_0 = 2$. Except for that, everything is as during your first visit.

b. What is the firm's marginal cost and average total cost now? Calculate, illustrate and discuss.
c. You argue that it is important to look at the labour input requirements to understand the differences in marginal costs during your first and second visit. Please explain to the management of Alpha Tech why this is the case.

6.3   Leontief & Sons Considers Expansion

Do you remember Leontief & Sons? We visited them in Exercise 5.3, and the question was then whether to move production of men's shirts to China. The brothers have decided to keep all three factories in the United States and are now interested in getting a better understanding of marginal costs. Refer to Exercise 5.3 for facts about their technology and let $w = r = 1$. Also assume that each factory has installed capital that is optimal for producing $Q = 1$.

a. What are the marginal costs of the Leontief brothers' shirt production? Calculate and illustrate. (Hint: start from the lowest marginal cost and move up, step by step).
b. One of the brothers says: "Let's expand production, and let's do it now!" Another brother is concerned about the costs of doing so. What do you think about this discussion?

6.4   Green Transition in the Cardboard Industry

A factory produces cardboard Q using the technology $Q = K_0^{0.5} L^{0.5}$, where K_0 is installed capital and L is labour. Wet pulp needs to be dried in the production process, and we assume that the electricity costs of doing so are proportional to production, $Z = zQ$, where z is the unit price of electricity. The cost function for a traditional ("brown") cardboard factory is given by:

$$C_B = wL + rK_0 + zQ$$

Assume the installed capital is given by $K_0 = 1$, the price of capital is $r = 1$, and the wage is $w = 1$.

The factory is now considering making a green transition: they want to invest in an incineration facility that produces heat from waste and residual materials. Such a plant costs $FC_G = 3$, and with this investment there is no longer any need to buy electricity, i.e., $Z = 0$. The cost function after a green transition then becomes:

$$C_G = wL + rK_0 + FC_G$$

a. What are the marginal costs and the total average costs (in the short run, i.e., with the installed capital fixed) before and after a green transition?
b. Which technology gives the lowest possible costs, and how is this affected by the price of electricity?

6.5   Conrad's Two Factories

Conrad has two factories; one located in the southern part of the county and the other further north. The factory in the South has costs $C_S = Q_S^2$, while the one in the North has costs $C_N = 0.6Q_N$. The time-horison is short-term, so think about these as variable cost functions. Conrad wants to keep total production at $Q = 1$ but is wondering what the optimal distribution of production should be between the two factories and has asked Anna and Brian for advice.

Use the "bathtub" approach shown in Chapter 6.4 in the textbook, where production in the South is measured from left to right and production in the North from right to left.

a. Brian suggests that the average costs in the two production facilities should be equalized. Calculate the costs with Brian's proposal and illustrate in a figure.
b. Conrad thinks it is most reasonable to split production evenly between the two factories. Calculate the costs with Conrad's proposal and illustrate in a figure. Whose suggestion is better: Conrad's or Brian's?
c. Anna disagrees with both. She says: "It is important to think marginally!" What does she mean by this? Calculate the costs with Anna's proposal and illustrate in a figure, where you also compare with the proposals from Brian and Conrad.

6.6   Tips

6.1 The Seamstress in Dar es Salaam
The point of this exercise is to explore the concept of opportunity cost: What is the value of the production inputs in their best alternative use? The task is open-ended in the sense that you need to think about what realistic alternative uses there are for Grace's time and her capital (production space and sewing machines).

6.2 Visiting Alpha Tech
This exercise is designed to help you become more familiar with calculating marginal costs for a firm with a balanced Cobb–Douglas production function, and Math Box 6.1 will be useful. The challenge with the exercise is to interpret what you

see, and in particular the changes in the cost functions from your first to your second visit.

6.3 Leontief & Sons Considers Expansion

From Exercise 5.3 we know that the optimal amount of capital to produce one unit of shirts ($Q = 1$) is given by $K = \alpha$: any amount of capital below this will make it impossible to produce one unit of shirts, while capital beyond this level is not needed.

To find marginal costs, we can start with variable costs, which are $VC = wL$. To express this in terms of units of output, Q, we need to use the production function:

$$Q = min\left(\frac{K}{\alpha}, \frac{L}{1-\alpha}\right)$$

For $Q < 1$, the installed capital is not binding, and production is determined by the input of labour, that is, $Q = L/(1 - \alpha)$. We can reorganize this and write it in terms of labour requirement as $L = (1 - \alpha)Q$. Plug this into the expression for variable costs, which gives you variable costs as a function of quantity Q. From there, take the derivative and you have marginal costs!

6.4 Green Shift in the Cardboard Industry

Investing in green energy brings additional fixed costs, but the benefit for the firm lies in reduced energy use and thus lower variable costs. Such an investment can also provide gains for society in the form of lower environmental emissions, but that is beyond the scope of this exercise.

In part (a), you are asked to calculate marginal costs and average total costs with brown and green technology. Math Box 6.1 gives you a recipe for how to compute these costs.

To solve part (b), start by finding the level of output that minimises average total costs. You can do this either by setting $(\partial ATC/\partial Q) = 0$ or by finding the output level Q where $MC = ATC$, both methods will yield the same answer. Then find the value of ATC for this output level; for the brown technology, this depends on the electricity price z. You can then find the value of z that yields the same minimum average total cost for the two technologies.

6.5 Conrad's Two Factories

The starting point is a situation where Conrad has two production units with different cost structures, and he wants to allocate a given level of output between them in a way that minimises costs. Various suggestions are on the table: Brian suggests allocating production such that average costs are equalised; Conrad thinks it makes sense for the two factories to produce equal amounts; while Anna focuses on marginal cost. The purpose of this exercise is to give you training in drawing a "bathtub" figure and using it to practice marginal thinking. When constructing the bathtub, you must mirror the production from the factory in the North and measure it from right to left. In this case, that's quite straightforward, since the cost function in the North is just a horizontal line!

I really like this exercise, I hope you enjoy it too!

Profit

7

Conrad on his way to the bank to discuss paper and profits

This chapter is about profits, and the central question is: How much should the firm produce? To answer this, the firm must consider not only costs but also the market price. The exercises give you practice in calculating the optimal level of production under different assumptions about prices and production technology, and in assessing whether production should be increased or reduced—or possibly shut down—in the short and long run.

© The Author(s), under exclusive license to Springer Nature Switzerland AG 2026   37
K. Bjorvatn, *Workbook for Microeconomics Made Simple*, Classroom Companion:
Economics, https://doi.org/10.1007/978-3-032-06357-1_7

7.1   Profits in the Green Cardboard Factory

The cardboard producer we met in Exercise 6.4 has now adopted the green technology, such that costs are given by $C_G = VC + FC_G = Q^2 + 4$. The firm operates
in a perfectly competitive market.

a. What level of production maximises profits if the price of cardboard is $P_1 = 4$?
   What is the firm's profit? Calculate and illustrate.
b. The price of cardboard now increases to $P_2 = 6$. Some argue that production
   should not be expanded, since it would increase costs. What would the profit
   be if production remains unchanged? Calculate and illustrate.
c. Do you agree that production should remain unchanged given the higher price?
   What is your suggestion?

7.2   Profits Go Up and Down

Use Fig. 7.1 in the textbook as your starting point. Assume a production function
$Q = K_0^{0.5} L^{0.5}$, with $K_0 = 1$, $r = w = 1$, and raw material costs zQ, where $z = 1$,
and let $P_2 = 5$.

   What are the quantities produced at points a, b, c and d in the figure? Round to
the nearest two decimal places.

7.3   More or Less Satisfied Capital Owners

This exercise builds on the previous one, but with a focus on the impact of price
on profits.

a. What are the profits for a profit-maximising producer when the price is $P_2 = 5$?
b. What is the lowest price the firm can tolerate before profits turn negative?

7.4   Higher Gas Prices: Should We Shut Down
the Factory?

Natural gas is an important source of energy in many industries. Using diagrams (no math required here), analyse how an increase in the price of natural
gas affects a firm's optimal production decisions and profitability (you can think
of a rise in gas prices as an increase in what we often refer to as z).

   When would you recommend shutting down the factory?

7.5   The Firm's Demand for Labour

Consider a firm with a balanced Cobb–Douglas production function $Q = K^{0.5}L^{0.5}$ and where labour is the only variable cost (i.e., ignore costs related to raw materials and energy, so $Z = 0$). Assume also that the installed capital is fixed at $K_0 = 1$.

a. What is its demand for labour (as a function of price and wage) in a profit-maximising firm?
b. Assume that the price is low, $P_{low} = 2$. Draw the labour demand curve in a figure with wage on the vertical axis and number of workers on the horizontal axis.
c. How many workers should the firm hire if the wage is high: $w_{high} = 1$? How about when it's low: $w_{low} = 0.5$?
d. Assume now that the price goes up to $P_{high} = 4$. How does the higher price affect the demand for labour? How many workers should the firm hire if the wage increases to $w_{high} = 1$?

7.6   Tips

7.1 Profits in the Green Cardboard Factory

This task is designed to give you a better understanding of profit maximisation and profitability. How much should a profit-maximising firm, which takes the market price as given, produce—and is this level of production profitable?

Math Box 6.1 gives you the method for calculating marginal cost, and the firm maximises profits by choosing a quantity such that marginal cost equals price. To determine whether the firm is profitable, you must compare the price with average total cost. What does it mean if profit turns out to be zero?

In part (b), you are asked to consider what happens if the price increases to $P_2 = 6$. One suggestion is that production should not be increased, since this would raise costs. But you know that profit maximisation implies marginal cost equals price, which leads to part (c) of the exercise. Compare the profits under unchanged production with your own proposal, where production is optimised given the new price.

7.2 Profits Go up and Down

From the discussion of Fig. 7.1 in the textbook, we know that point a represents a situation where profits are zero because average total cost is very high (driven by fixed costs); then profit increases as the quantity produced rises, reaching point b, which is the profit-maximising quantity; profits then decline beyond this point, reaching point c where it is zero again—this time due to high variable costs; and finally, point d, where operating profits are zero.

In this exercise, you will use a specific production function and a numerical example to identify the levels of production that correspond to these four points in the figure.

Math Box 7.1 discusses profitability for a similar production function and a similar numerical example. Insert the price $P = 5$ into the relevant expressions, and you'll find your answers!

7.3 More or Less Satisfied Capital Owners

From Math Box 7.1, we know that profits are given by $\pi = PQ - C = PQ - (Q^2 + Q + 1)$.

Initially, the price is $P = 5$, and in the previous exercise we found that the producer then chooses $Q = 2$. Insert this into the profit expression to easily find the level of profits.

To answer part (b), we need to find the quantity that gives the lowest possible average total cost, ATC, defined as:

$$ATC = \frac{C}{Q}$$

With $C = Q^2 + Q + 1$, we then have:

$$ATC = \frac{Q^2 + Q + 1}{Q} = Q + 1 + \frac{1}{Q}$$

The quantity that gives the lowest possible ATC can be found by differentiating this expression and setting it equal to zero:

$$\frac{\partial ATC}{\partial Q} = 0$$

We also know that MC crosses ATC at its minimum point. Since $MC = 2Q + 1$, you can use the value of Q that gives the lowest cost to find the lowest price the firm can tolerate before profits turn negative.

7.4 Higher Gas Prices: Should We Shut Down the Factory?

This exercise challenges you to consider how an increase in the price of energy (z) affects the firm's production choice and profitability. Remember that an increase in z shifts both the MC curve and the ATC curve upward. At some point, the gas price becomes so high that even at the optimal choice of output (choosing quantity such that marginal cost equals price), profits become zero. A gas price above this critical level will necessarily lead to negative profits.

7.5 The Firm's Demand for Labour

The demand for labour by a profit-maximising firm is given by the condition $PMP_L = w$, that is, where the marginal revenue product of labour equals the wage—the income generated by the last worker hired must equal what he or she costs. The marginal product of labour from a Cobb–Douglas production function should be familiar (see

Math Box 5.1), and in part (a) you will use this to derive labour demand as a function of the wage and the price of the product.

In parts b and c, use the result from part (a) to draw the labour demand curve in a diagram with the wage on the vertical axis and the number of workers on the horizontal axis.

Note the difference between labour requirement (see Exercise 6.2) and labour demand, which is the focus here. Labour requirement is the amount of labour needed to produce a given quantity and is illustrated in a diagram with L on the vertical axis and Q on the horizontal. Labour demand, by contrast, is derived from profit maximisation and is illustrated in a diagram with w on the vertical axis and L on the horizontal axis.

Perfect Competition

8

Anna points out that trade may create new opportunities for The Mill

This chapter explores what determines the price in a market—more specifically, in a perfectly competitive market. Working with these exercises will give you a deeper understanding of how markets are affected by shocks to supply and demand, and of opening up to international trade.

Many of the exercises deal with electricity and energy markets, for two reasons: First, these are homogeneous goods (electricity is electricity, gas is gas), and therefore well suited for analysis under perfect competition. Second, energy markets are high on the international political agenda, tied to fundamental issues of climate and conflict, and therefore interesting from a policy perspective.

© The Author(s), under exclusive license to Springer Nature Switzerland AG 2026
K. Bjorvatn, *Workbook for Microeconomics Made Simple*, Classroom Companion:
Economics, https://doi.org/10.1007/978-3-032-06357-1_8

8.1 Demand for Gas from Households and Industry

Gas is an important source of energy in Europe, both for households and for industry. Assume that households have the demand function $Q_H^D = 5 - P$ while industry has the demand function $Q_I^D = 6 - 2P$.

a. What is the choke price for the two consumer groups (households and industry)?
b. Illustrate in a diagram the demand from the two consumer groups separately and combined.
c. What happens to the demand from households and industry if the gas price rises from a low price $P_l = 1$ to a high price $P_h = 2$?

8.2 War and Gas Prices in the Short and Long Run

In Exercise 8.1, we treated the international gas price as given. In this exercise, we will consider what happens to the price of gas when there is a negative supply shock in Europe, caused by reduced access to gas from Russia in the wake of the Ukraine war.

We examine two scenarios. The first is short term, where consumers have little opportunity to adjust their energy consumption. The second is longer term, where consumers have had time to invest in alternative energy sources, such as solar panels, and in energy-saving measures, such as improved insulation.

We aggregate the different consumer groups and assume that gas demand is given by:

$$Q^D = \alpha - \beta P$$

In the short run, $\alpha_{short} = \beta_{short} = 1$, while in the long run $\alpha_{long} = 1.5, \beta_{long} = 2$.

We can think of this as a combination of a general increase in the demand for gas over time (higher α) and investments that make it easier to switch to other energy sources, such as solar or wind, if the price of gas becomes high (higher β).

a. Draw the demand curve in both the short and the long run. What is the choke price for consumers in the short run and in the long run?
b. The gas supply is given by $Q^S = -\gamma + P$. Assume that $\gamma = 0$ to begin with. Calculate the market equilibrium in the short and long run and illustrate. What is the price elasticity of demand in the short- and long-run equilibrium?
c. With the invasion of Ukraine and the breakdown of trade relations with Russia, we observe a shift in the supply curve, now given by $\gamma' = 0.3$. What happens to the market equilibrium in the short run? And what about in the longer run? Comment on your findings.

8.3   Pennies from Heaven

In Norway, most electricity production comes from hydropower, and many people wonder why electricity is so expensive here. There's plenty of rain, it's like pennies from heaven! At zero cost, many would argue that electricity should be cheap, perhaps even for free!

Assume that production of hydropower is described by a Leontief technology $Q^S = \min(V, K)$, where V is water ("vann" in Norwegian ☺) in the reservoirs and K is the installed hydropower capacity, determined by the size of the dams and turbines.

We assume that capacity is fixed at $K = 1$. The water level, however, depends on rainfall, and we will study both a dry year, with $V_l = 0.75$, and a wet year, with $V_h = 1.25$.

We consider two demand scenarios: summer and winter. In the summer, demand is low and given by $Q_l^D = 0.5 - P$, while in the winter, demand is high $Q_h^D = 1.5 - P$.

a. What will the electricity price be in summer and winter during the dry year?
b. What will the electricity price be in summer and winter during the wet year?

8.4   Climate Change and International Trade in Agriculture

Demand for agricultural products in your home country is given by $Q^D = 100 - P$. Climate change is making agricultural production increasingly vulnerable to weather variability. Consider three weather scenarios:

Scenario 1:   A normal year, where the supply of agricultural products is given by $Q_{norm}^S = -20 + P$

Scenario 2:   A bad year, where a climate shock has reduced yields, such that supply is $Q_{bad}^S = -40 + P$

Scenario 3:   A good year, where plentiful sun and rain lead to a supply of $Q_{good}^S = P$

a. Consider first a situation without international trade (that is, autarky). What is the market solution (price and quantity) in the three scenarios? Calculate and illustrate.
b. Now assume that your country has opened up for international trade in agricultural goods, and that the international price is $P_{int} = 60$ (which we assume is unaffected by weather conditions in your country). What is the market solution (production, consumption, and trade) in the three scenarios? Calculate and illustrate.

c. Compare the situation with and without international trade. Who gains and who loses from trade in each of the weather scenarios—consumers or producers? Think about an increase in quantity consumed as a gain for consumers and an increase in quantity produced as a gain for producers (there will be more about consumer surplus and producer surplus in the next chapter).

8.5   Anna Points Out That Trade Can Create New Growth Opportunities for the Mill

Conrad is worried that international trade in paper will create challenges for the paper mill. But Anna shows that trade can also create opportunities for increased production. She outlines three scenarios and, so as not to bore her grandfather, relies solely on graphical analysis—no math involved.

In each scenario, she compares the outcome under autarky with what the equilibrium would be under international trade and explains why trade can be good for business. To illustrate her points clearly, in each scenario she starts from a situation where the international price equals the autarky price.

Scenario 1:   A negative shift in the demand curve, for example due to worse economic conditions at home.

Scenario 2:   A positive shift in the supply curve, for example due to lower raw material prices that reduce marginal costs.

Scenario 3:   A higher international price, for example due to increased demand internationally.

8.6   Tips

8.1 Demand for Gas from Households and Industry
In this exercise, you will become more familiar with the demand curve from different types of consumers and how to aggregate these to construct the market demand curve. The context is the gas market, where the price sensitivity of demand differs between households and industry.

Total demand is found by summing the two demand functions $Q^D = Q_H^D + Q_I^D$.

Note that the aggregate demand curve has a kink at the choke price for industry (P_I^{choke}). For prices above this level, only households want to buy gas.

8.2 War and Gas Prices in the Short and Long Run

In part (a), you will need the definition of the choke price:

$$P_D^{choke} = \frac{\alpha}{\beta}$$

In part (b), the expression for the equilibrium price given in Math Box 8.3 will be useful:

$$P^* = \frac{\alpha + \gamma}{\beta + \delta} \text{ Equilibrium price}$$

You find the price in the short and long run by inserting the demand parameters provided in the exercise, and from that you can determine the corresponding quantities.

In part (c), use the formula for the equilibrium price P^* shown above and insert the relevant demand parameters along with the new value $\gamma' = 0.3$.

8.3 Pennies from Heaven

In this exercise, you are introduced to hydropower producers who use Leontief technology to produce electricity, with water and capital as inputs. With this technology, the supply curve is horizontal up to the capacity limit, and then vertical. Since water is free, the marginal cost is zero up to the capacity limit. The scarcest factor determines the production capacity.

With the numbers given in the exercise, you will see that in a dry year, water availability defines the capacity limit, while in a wet year the installed capital (dams, turbines) is the limiting factor.

You are presented with different demand scenarios for summer and winter. In each case, it can be useful to check whether demand uses up the entire capacity or not. You can test this by setting the price equal to zero and seeing what demand is. If the entire capacity is not used, then marginal costs define the price (which, with free water, is zero). If demand requires full use of capacity, the price rises to clear the market. The market price will be the one that makes demand equal to supply (as defined by the capacity limit).

High demand in winter can cause electricity to cost much more than water!

8.4 Climate Change and International Trade in Agriculture

Without trade (autarky), market equilibrium is found where supply intersects demand. We find the market equilibrium as $Q_i^S = Q^D$, where $i = normal, bad, good$ year.

With international trade and a price $P_{int} = 60$, we can find supply in a normal year by using $P = P_{int} = 60$ in the supply function $Q_{norm}^S = -20 + P_{int}$.

If the autarky price is higher than the international price, we get imports; if lower, exports.

8.5 Anna Points Out that Trade Can Create New Opportunities for Growth for the Mill

In this exercise, you will use the market model with international trade to describe how different shocks affect production for Conrad and the other paper producers. This is a drawing exercise, no math involved, so technically it should not be too demanding. But it offers a good opportunity to think about markets with international trade.

Economic Efficiency

9

Externalities, efficiency, and engagement!

This chapter discusses the properties of market equilibrium, focusing on economic efficiency, but also considering distributional aspects. We study how the market creates surplus for consumers and producers, how disturbances in the price mechanism generate economic inefficiency, and how externalties give rise to deadweight loss.

The exercises invite you to apply the theory to important topics such as minimum wage, environment, and international trade.

© The Author(s), under exclusive license to Springer Nature Switzerland AG 2026
K. Bjorvatn, *Workbook for Microeconomics Made Simple*, Classroom Companion:
Economics, https://doi.org/10.1007/978-3-032-06357-1_9

9.1   War on the Continent, Higher Gas Prices and Economic Surplus

Conrad reads in the newspaper about the war in Ukraine and that the loss of Russian gas deliveries has caused the price of natural gas to go up. He thinks that this price increase cannot be economically efficient.

Imagine a situation where demand for gas is given by:

$$Q^D = 1 - P$$

While supply is given by:

$$Q^S = -\gamma + P$$

Before the war in Ukraine, $\gamma_{peace} = 0$, but with the halt in gas deliveries from Russia, we get a shift in the supply curve given by $\gamma_{war} = 0.2$.

a. What is the equilibrium and the economic surplus in this market before the war?
b. How does the war affect the equilibrium and the economic surplus? Do you agree with Conrad that the war has led to a loss in economic efficiency?
c. Conrad thinks that when the war is over and we return to $\gamma_{peace} = 0$, the economic surplus will increase, and the economy become more efficient. Is he right?

9.2   The Labour Market and Minimum Wage

Market theory is flexible and can be used to study many types of markets, including the labour market. Consider the figure below, where the wage (w) is on the vertical axis and the number of workers (L) on the horizontal axis.

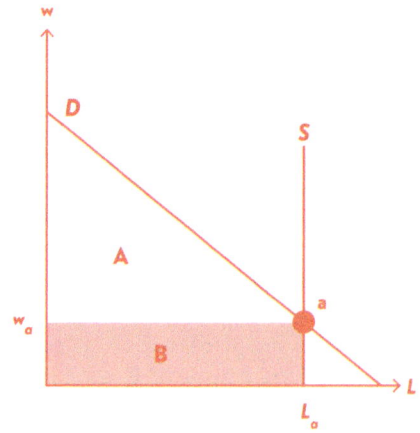

The demand for labour is given by D: higher wages make it less profitable to hire people (see textbook Chapter 7.5 on firms' demand for labour).

For simplicity, we let the supply of labour S be vertical (as you might remember from Chapter 3, a wage increase creates both a substitution effect pushing towards more work, and an income effect pushing towards less work, so a vertical supply curve is not completely unreasonable).

Market equilibrium is given by wage w_a where the supply curve intersects the demand curve, with L_a employees.

a. How would you interpret areas A and B in this figure?
b. Imagine the government thinks the market wage is unreasonably low and decides to introduce a minimum wage w_{min}. How will this affect the market outcome?

9.3   The Price of Cheap Petrol

Many oil-rich countries (such as the Gulf states) sell petrol at a very low price to domestic consumers. Assume that the government controls both petrol production and sales, setting the domestic price equal to marginal cost (in practice, the autarky price), while exporting the remaining supply at the international price. Use graphical tools (no math here!) to argue why such a low-price policy is bad for economic efficiency.

9.4   When the Invisible Hand Fails: Deadweight Loss from Pollution

Look at Fig. 9.4 in the textbook. Assume demand is given by $Q^D = 5 - P$ and supply by $Q^S = -1 + 2P$. There is an environmental cost $e = 1.5$ per unit produced.

a. Calculate price and quantity in the market equilibrium (point a) and in the economically efficient solution (point b).
b. How large is the deadweight loss (†) in the market equilibrium?

9.5   Are Imports a Threat to Innovation and Economic Efficiency?

Imagine an industry where supply is given by $Q^S = -2 + P$, which in inverse form can be expressed as $P = 2 + Q^S$. This represents the firms' marginal costs.

Assume that production creates innovation, which adds to the knowledge capital in society. This is an example of a positive externality. The social marginal cost is therefore given by $P = 2 + Q^S - h$, where h is the human capital externality (which we assume is proportional to production). Let $h = 2$, in which case the social marginal cost simplifies to $P = Q^S$.

Let demand be given by $Q^D = 10 - P$, and assume the market is open to international trade with an international price $P_{int} = 4$.

a. What is the market solution? Calculate and illustrate.
b. What is the deadweight loss in the market equilibrium, and what is the economically efficient production level? Calculate, illustrate, and comment on whether import is a threat to innovation and economic efficiency in this case.

9.6   Tips

9.1 War on the Continent, Higher Gas Prices and Economic Surplus

This exercise shows how a supply shock affects the total economic surplus (consisting of consumer surplus and producer surplus). Technically, the exercise shouldn't be too demanding, but it gives you practice in calculating the areas that make up the different surpluses, and in reflecting on what the concept of market efficiency entails.

As you know, consumer surplus is the area under the demand curve and above the price, and this is a triangle. Find the consumers' choke price and then use the equilibrium price and quantity to calculate the area of the triangle. Producer surplus is the area above the supply curve and below the price. Again, this is a triangle, so find the producer's choke price and then calculate the area.

When you in part (b) assess whether the war creates an efficiency loss, you must remember the criterion for economic efficiency: marginal cost equals marginal willingness to pay. A higher price is not necessarily a sign that the market is functioning poorly!

9.2 Trade Unions and Minimum Wage

The analysis of the labour market is completely parallel to the analysis of a market for any other good, the only difference being that we have the price of labour—that is, the wage—on the vertical axis (instead of the price of a product), and the number of workers on the horizontal axis (rather than quantity produced).

With a minimum wage, employers will demand fewer workers, and the supply of labour at that wage will be higher than the demand, i.e., unemployment. There is also an efficiency loss, defined as the difference between what firms are willing to pay for labour and the wage that workers are willing to work for.

9.3 The Price of Cheap Petrol

In a resource rich country, why should local consumers pay the same price as consumers abroad? This exercise shows why economists are not in favour of handing out cheap petrol to the population.

The task is based on graphical analysis and therefore shouldn't be technically complicated, but the insight is important. Cheaper petrol for the people may increase consumer surplus, but it creates a loss in overall economic efficiency.

A tip: the low-price policy implies that producers will supply a quantity to domestic consumers corresponding to the autarky equilibrium, while total production is determined by the point where the supply curve intersects the international price. The difference between total production and the quantity sold locally is exports. And the local price (given by the autarky price) is lower than the export price—that's the whole point of the low-price policy.

9.4 When the Invisible Hand Fails: The Deadweight Loss from Pollution

You can find the firm's supply curve S by rewriting the supply function $Q^S = -1+2P$ into its inverse form, expressing price as a function of quantity $P = 0.5 + 0.5Q^S$. This gives an expression for the firm's marginal cost.

The social marginal cost $S + e$ is then $P = 0.5 + 0.5Q^S + e$. Furthermore, in equilibrium we will have $Q^D = Q^S$, so, with this, you basically have all the information you need to get started!

9.5 Are Imports a Threat to Innovation and Economic Efficiency?

In this exercise, we combine the topics of externalities and international trade.

We find the producer equilibrium by inserting the international price P_{int} into the producers' supply function $Q^S = -2 + P$, and similarly, we find consumption by inserting the international price into the demand function $Q^D = 10 - P$. You will see that in this scenario there is import, since with the given international price, $Q^D > Q^S$.

Given that production generates a positive learning effect for society, the social marginal cost is lower than the private marginal cost, and the socially optimal solution is found where the social marginal cost curve intersects the international price line P_{int}.

Economic Policy

10

"New tax on paper". Conrad is worried about the factory's profitability

This chapter discusses how economic policy—particularly taxation—affects the market. The exercises invite you to assess the consequences the consequences of government interventions for consumer surplus, producer surplus, and economic efficiency, dealing with problems related to environmental policy and how price or quantity regulation affects the market.

© The Author(s), under exclusive license to Springer Nature Switzerland AG 2026
K. Bjorvatn, *Workbook for Microeconomics Made Simple*, Classroom Companion:
Economics, https://doi.org/10.1007/978-3-032-06357-1_10

10.1   Lower Electricity Tax, Same Bill!

High electricity prices have led to a popular demand to remove the electricity tax. One might assume that lowering this tax would result in a corresponding drop in the electricity price for consumers, but it's not that simple.

Assume that electricity production is associated with zero marginal cost (cheap hydropower, solar, wind, and nuclear power) up to a capacity limit of $Q^S = 1$, and then a vertical supply curve at this point. Let the demand for electricity be given by $Q^D = 1.5 - P^D$, where P^D is the price consumers pay.

a. What is the equilibrium price and quantity without an electricity tax?
b. What happens if an electricity tax $t = 0.5$ is introduced? Study the effect on the consumer price (P^D) and the producer price (P^S), where $P^D = P^S + t$. Who bears the tax burden?
c. What happens if the government gives in to popular pressure, and removes the electricity tax? Do you think consumers will be pleased?

10.2   Subsidies, Incidence, and Deadweight Loss

Consider a market where demand is given by $Q^D = 100 - P$ and supply by $Q^S = -20 + P$.

a. What is equilibrium price and quantity in this market?
b. The Producer Party has called for a subsidy to producers of $s = 20$ for each unit produced. This will create a wedge between the consumer price and the producer price, so that $P^D + s = P^S$. What is the equilibrium with such a subsidy?
c. Do you think the Producer Party will be satisfied with this policy? How about the Efficiency Party?

10.3   The Environmentalists

Consider a market with demand $Q = 100 - P$ and where producers have a constant marginal cost $MC = 50$. There's an environmental cost $e = 10$ per unit produced, which the firms do not take into account when making their production decisions.

a. What is the market outcome, the deadweight loss, and the total economic surplus without any government intervention? Calculate and illustrate.
b. What is the total economic surplus with an optimal environmental tax? Calculate and illustrate.
c. The Green Party proposes a tax of $t = 20$ on production. What are the economic consequences of such a tax? Calculate the price and quantity produced, the deadweight loss, and the total economic surplus, and illustrate. Compare the

Green Party's proposal with the market outcome you found in part (a) and discuss your observations.

10.4 Trade Policy for Innovation?

This question builds on Exercise 9.5 in the previous chapter, where you analyzed positive externalities related to innovation. There we saw that the market solution leads to too much import and too little domestic production. Does this mean we should ban imports?

Again, consider a market where supply is given by $Q^S = -2 + P$ and demand by $Q^D = 10 - P$. The international price of the product is $P_{int} = 4$, and there is a positive knowledge externality $h = 2$ for each unit produced, such that the value of the human capital to society is $H = hQ$.

a. Some propose banning imports, which here means introducing autarky. What is the total economic surplus in this case?
b. Another proposal is to allow imports but introduce a production subsidy s per unit equal to the knowledge externality h: that is, $s = h$ What is the total economic surplus in this case?
c. Compare the outcomes of the two proposals, autarky and production subsidy. Which policy would you recommend?

10.5 Odd–Even Driving vs. Road Tolls

Many large cities are struggling with congestion, and various policies have been suggested to deal with the problem. One interesting proposal is odd–even driving, a scheme where cars with license plates ending in even numbers are allowed in on one day, and those with odd numbers on the other. Since half of the cars have plates ending in even numbers and half in odd numbers, this would automatically halve the traffic into the city centre.

Others argue that a toll, where commuters pay a tax for entering the city, is a better idea to achieve the same reduction in traffic.

The willingness to pay for driving into the city is given by $P = 1 - Q$, and assume that there are no other costs associated with driving into the city. Therefore, without any intervention from the municipality, everyone would drive, $Q = 1$ (think of "1" as meaning "100%" of drivers). Which policy would you recommend: odd-even driving or the road toll, where the size of the toll is such that traffic is cut in half?

10.6   Tips

10.1 Lower Electricity Tax, Same Bill!

This exercise invites you to think about tax incidence: Who bears the cost of a tax (or benefits when a tax is removed or a subsidy introduced)? This is an important issue, and the market model gives answers that are not always immediately obvious. Theory is useful! The task is linked to Fig. 10.2 in the textbook, so it could be a good idea to review the discussion there.

The application is the electricity market, where producers supply power at zero marginal cost up to a capacity limit given by the installed capital. This means the supply curve is perfectly elastic (constant marginal cost) when demand is low, and perfectly inelastic (vertical supply curve) when demand is sufficiently high. The exercise contains a lot of information, so really there is not much more to say than: good luck!

10.2 Subsidies, Incidence and Deadweight Loss

Most of the analysis in Chapter 10 deals with taxes, but here you will look at the effect of a subsidy. There is a clear connection between the analysis of tax incidence, as described in Math Box 10.1, and the effect of a subsidy.

A tax creates a wedge $P^D = P^S + t$, while a subsidy creates a wedge $P^D + s = P^S$. We see that the subsidy can be interpreted as a negative tax: $s = -t$.

Part (a) deals with the situation before the subsidy, which should not pose major problems. In part (b), when you introduce the production subsidy, remember that the equilibrium condition $Q^D = Q^S$ can now be written as:

$$100 - P^D = -20 + P^S$$

You can then use $P^D + s = P^S$ in this expression and find:

$$100 - P^D = -20 + \left(P^D + s\right)$$

Then you have the consumer price, and based on this, the producer price, and you're on your way!

10.3 The Environmentalists

This task is not particularly difficult technically, but it invites you to think about how a tax can create environmental benefits, while these benefits must also be balanced against other considerations, such as consumer interests. The optimal environmental tax is a trade-off between these different concerns. Remember to include tax revenues in the economic surplus when relevant.

10.4 Trade Policy for Innovation?

This exercise has an important message: Policy should be as targeted as possible to avoid unintended side effects. The application is on positive spillover effects from innovation.

In part (a), calculate the economic surplus under autarky, which should not be too difficult. In part (b), look at a production subsidy as an alternative to autarky. Find the market solution with subsidy $s = h = 2$ and illustrate the economic surplus, consisting of consumer surplus, producer surplus (with subsidy), the value of human capital $H = hQ$, and remember to subtract the size of the subsidy. You will see that economic surplus is higher with the production subsidy than with autarky. Try to identify the increased surplus in the figure. In part (c), reflect on why the production subsidy is better than autarky.

10.5 Odd–Even Driving vs. Road Tolls

This exercise will challenge you to think about the most reasonable intervention from an economic efficiency perspective. The application is very relevant: Authorities want to limit car traffic into large cities and have proposed so-called odd–even driving, where only half the drivers are allowed in on a given day.

The demand to use a car into the city is given by $Q = 1 - P$, and with zero marginal cost, everyone with a positive willingness to pay would drive. The entire area under the demand curve is then consumer surplus, which in this cases constitutes the economic surplus. Odd–even driving reduces demand to $Q' = 0.5(1 - P)$. The demand curve will then cross the horizontal axis at $P = 0 \Rightarrow Q' = 0.5$, while the choke price remains the same. With the situation without government intervention as reference, how much consumer surplus is lost?

In part (b), consider an alternative policy, namely a road toll. What price per passage P (which can think of as the toll) halves the traffic, i.e., $Q = 0.5$? The economic surplus now consists of both consumer surplus and the toll revenues, $T = PQ$. In part (c), compare the efficiency loss of the two interventions. In both cases, car usage is halved, so why would you recommend the toll ? The answer lies in who uses the car…

Monopoly

11

Conrad is wondering how to price his textbooks

So far in the book, we've studied firms without market power—they are small and take the market price as given. But in this chapter, we look at a situation where there is only one firm in the market: a monopoly.

A monopolist recognises that there is a relationship between the quantity produced and the price they can charge for the product. The exercises in this chapter will help you better understand the factors that determine the monopolist's choice of price and quantity—and the consequences of monopoly power for society.

© The Author(s), under exclusive license to Springer Nature Switzerland AG 2026
K. Bjorvatn, *Workbook for Microeconomics Made Simple*, Classroom Companion:
Economics, https://doi.org/10.1007/978-3-032-06357-1_11

11.1   "Green Is Good"—But at What Price?

Alpha Books is about to publish a new vegetarian cookbook called *Green is Good*. Anna is advising Conrad on what price the publisher should set for the book (since they are the only one publishing this book, Alpha Books is a monopolist), and she looks to the neighbouring country for inspiration, where a similar book exists: "Very Veggie". Its publisher—also a monopolist—has conducted a market analysis that she thinks might be useful to consider.

a. The market analysis shows that when the price of Very Veggie increased by 10 percent, it led to a 5 percent drop in demand. That is, the price elasticity of demand was $\varepsilon = 0.5$. Anna is not convinced that the publisher chose the optimal price. What do you think?

b. Anna assumes that demand in the two countries is the same, given by $Q = \alpha - \beta P$. The price of Very Veggie is $P = 1/3$, and the quantity sold is $Q = 2/3$. From the market study, she also knows that $\varepsilon = 0.5$. Can you help her find α and β in the linear demand function?

c. Conrad tells Anna that the marginal cost of producing Green is Good is constant and given by $MC = 0.2$. What price should Conrad choose to maximise profits?

d. What is the price elasticity of demand at the profit-maximising quantity?

11.2   The Monopolist on the Mountain

Anna is a hiking enthusiast and often climbs one of the mountains near the School of Economics. At the top of the mountain, there's a kiosk—and it's a long way to the next one. She's thirsty and buys a bottle of soda. And gets annoyed. Six euros for a soda?!

"How many bottles do you even manage to sell at such a high price?" she asks the owner.

"Four bottles a day, actually", he replies.

The kiosk owner claims the high price is due to high costs—he says it costs four euros to purchase and transport each bottle of soda up the mountain. But Anna is sceptical.

a. What is the Lerner index based on the kiosk owner's information? And based on this information, what must the price elasticity of demand be, given that he is a profit-maximising monopolist?

b. Anna conducts a quick survey among the other hikers and finds that a 10 percent increase in price would lead to a 15 percent decrease in demand. She confronts the kiosk owner with this finding and claims that he has probably exaggerated his costs. How can she say that?

c. Back from her hike, she meets Brian at a café, tells him about the conversation with the kiosk owner, and tries to convince her friend that she's right. She starts from a linear demand function $Q = \alpha - \beta P$ and draws two diagrams: one based on the kiosk owner's information, and one based on her own survey. Find α and β in both cases and illustrate.

11.3   One Monopolist, Two Markets, Two Cost Scenarios

A monopolist sells a product in two separate markets. In market A, demand is given by:

$$Q_A = 2 - 2P$$

And in market B:

$$Q_B = 1 - P$$

Consider two cost scenarios: In the first, $TC = (1/3)Q$, while in the second, $TC = (1/2)Q^2$.

What price and quantity will the monopolist choose in the two markets in the two cost scenarios? Comment on the differences that you observe.

11.4   Efficiency Loss Under Monopoly

Consider the same monopolist as in the previous exercise, and let's focus on the market where demand is $Q = 1 - P$. Again, there are two cost scenarios: $TC = (1/3)Q$ and $TC = (1/2)Q^2$. We shall refer to these as cost scenarios A and B, respectively.

What is the deadweight loss from monopoly (compared to the perfectly competitive solution)?

Calculate and illustrate in a diagram. Comment on the difference in efficiency loss between the two cost scenarios.

11.5   Natural Monopoly and Economic Surplus

A monopolist has marginal cost $MC = 0$ and a fixed cost $FC = 10$. Demand is given by $Q = 10 - P$.

a. What is the monopoly outcome? What is the profit, and what is the total economic surplus (that is, the sum of producer and consumer surplus)?
b. Suppose the government takes ownership of the firm with the goal of maximising total economic surplus. What price will the government choose, and what is the efficiency gain compared to the monopoly solution you found in part (a)?

c. An alternative approach is that the government takes ownership and instructs the management to set the lowest possible price without making a loss (i.e., zero profits). Discuss this solution in relation what you found in part (b).

11.6   Tips

11.1 "Green Is Good", But at What Cost?

To solve part (a), it may be helpful to look at Fig. 11.2 in the textbook. We know that, to maximise profits, a monopolist should operate where demand is elastic. If a price increase of 10 percent leads to a fall in demand of 5 percent, what does that tell us about the price elasticity of demand?

As for part (b), note that we can find the price elasticity of demand for the linear demand function $Q = \alpha - \beta P$, on inverse form $P = (\alpha - Q)/\beta$, as:

$$\varepsilon = -\frac{\partial Q}{\partial P}\frac{P}{Q} = \beta\frac{\alpha - Q}{\beta Q} = \frac{\alpha - Q}{Q}$$

You can then use the information provided from the market survey to find α and β in the demand function.

The rest of the exercise should be straightforward. With the demand function you calculated in part (b), and the marginal cost from Conrad, you can easily find the optimal price and optimal quantity. With this optimal solution, you can then (to solve part d) use the elasticity formula above to calculate the elasticity at Conrad's optimal choice.

11.2 The Monopolist on the Mountain

From chapter 11.4 in the textbook, we know that the Lerner index for a monopolist is:

$$\frac{P - MC}{P} = \frac{1}{\varepsilon}$$

This formula is useful because it shows the relationship between the price markup and the price elasticity of demand. From the claimed price markup, you can find the elasticity ε that must hold for the profit-maximising kiosk owner. Anna conducts her own market survey and finds a much lower estimate of the price elasticity of demand, so she suspects that the marginal cost is much lower than the kiosk owner claims.

In part (c), you can use the midprice rule (see Math Box 11.1):

$$P_{Mon} = \frac{1}{2}\left(\frac{\alpha}{\beta} + c\right)$$

This is useful to find the relationship between the monopoly price and the demand parameters in the linear demand function $Q = \alpha - \beta P$, both for the kiosk owner's claimed marginal cost $c = 4$ and Anna's estimate. Remember also that the kiosk sells four sodas per day, $Q = 4$.

11.3 One Monopolist, Two Markets, Two Cost Scenarios

The goal of this exercise is to study how a monopolist's price is affected by differences on the demand side (market A and B) and on the cost side.

Note that $TC = (1/3)Q$ implies constant marginal cost $c = (1/3)$. Here you can use the midprice rule to find the monopolist's price.

Total costs $TC = (1/2)Q^2$ implies increasing marginal costs $c = Q$. Here, you need to use the first-order condition for the monopolist's choice of quantity, given by marginal revenue equals marginal cost, and then find the price corresponding to that quantity. But the procedure is well known, so technically this exercise should not pose too many problems.

11.4 Deadweight Loss from Monopoly

You can use the information for Market B from the previous exercise, where you found the monopolist's price and quantity in the two cost scenarios. But here the focus is on the deadweight loss caused by the monopoly.

It may be helpful to draw graphs of the markets and use these as guides for your calculations. Especially when it comes to finding the area of the deadweight loss triangles, having these figures in front of you is very useful.

Good luck!

11.5 Natural Monopoly and Social Economic Surplus

The goal of this exercise is to become better acquainted with natural monopolies, a situation where average costs decline as production increases. This is usually due to large fixed costs and low (often constant) marginal costs.

For part (a), note that since the monopolist has constant marginal costs (which are zero), you can use the midpoint rule to find the monopolist's optimal price:

$$P_{Mon} = \frac{1}{2}\left(P_D^{choke} + c\right)$$

With the demand function $Q = 10 - P$, we know the choke price $P_D^{choke} = 10$, and with zero marginal cost $c = 0$, you can easily find the monopoly price.

The total economic surplus is given by consumer surplus plus producer surplus. Producer surplus is the same as operating profits, that is, revenue minus variable costs. Since the variable costs here are zero, producer surplus simply equals revenue.

In part (b), find the socially optimal solution, which is given by price equal to marginal cost. Since marginal cost is zero, the price should also be zero.

In part (c), you are told the firm must set a price that yields zero profit, meaning a price where revenue exactly covers total costs. Since variable costs are zero, revenue must cover fixed costs, $FC = 10$. This implies a price equal to average total cost. Here, you will need to use the quadratic formula, as the optimal quantity results from solving a quadratic equation.

Round your answer to two decimals and assume the firm chooses the solution closest to the socially optimal solution described in part (b).

Oligopoly

12

Conrad and the competitor quarrel about cardboard

This chapter analyses the interaction between firms with market power—a situation we call *oligopoly*. Oligopoly theory consists of a range of specialised models, and the exercises explore some variations of the material covered in the textbook. What happens if we change the number of firms in the market, or introduce differences in costs or demand? Can advertising your product also benefit your competitor? And what are the implications for economic efficiency if a foreign company enters the market? These exercises offer a deeper understanding of the theory by applying it to a range of interesting questions.

© The Author(s), under exclusive license to Springer Nature Switzerland AG 2026
K. Bjorvatn, *Workbook for Microeconomics Made Simple*, Classroom Companion:
Economics, https://doi.org/10.1007/978-3-032-06357-1_12

12.1   Is a Merger Profitable?

A merger means two firms joining forces to form a single decision-making unit. Assume they split the profits equally after the merger—like a marriage between equals.

a. Initially, there are two firms in the market, Firm A and Firm B. They have identical constant marginal costs, and for simplicity we assume that $c = 0$ (so revenue and profits are the same). Quantity is the strategic variable, meaning that the market structure is of the Cournot type. Market demand is given by $Q = 1 - P$. What is the gain from the merger (that is, the shift from duopoly to monopoly) for the two firms in this case?

b. Now assume that there are not two, but three firms in the market to begin with—a triopoly consisting of Firm A, B, and C. All three firms are identical, with the same constant marginal cost $c = 0$ and quantity as strategic variable. How are profits affected by a merger between A and B in this case, changing the market structure from a triopoly to a duopoly between the merged unit AB and the independent Firm C?

12.2   Foreign Ownership, Competition and Economic Efficiency

Imagine a market with demand given by $Q = 1 - P$. Initially, there is only one producer in this market: a domestically owned firm, Firm A, with zero marginal cost ($c = 0$).

a. What is the monopolist's optimal choice of quantity and price? What is the consumer and producer surplus in this case? Calculate and illustrate.

b. A foreign producer, Firm B, then enters the home market. This firm produces the same product as Firm A, also with zero marginal costs. The competition is of the Cournot type (i.e., quantity competition). What are the economic consequences of this entry, if producer surplus goes to the foreign owners and therefore does not count as part of home country's economic surplus? Calculate and illustrate.

c. Now assume that the home firm is a Stackelberg leader, with the foreign firm following. What are the consequences of the foreign entry in this case? Compare with your findings in part (b).

12.3   Lower Costs, Higher Food Prices?

You've been hired as an economist at the Competition Authority and are tasked with studying the grocery retail market. The market consists of two chains: A and B. The Competition Authority has uncovered large differences in the chains' purchasing prices—firm A has managed to secure lower prices on cheese, eggs, and a range of other goods. Your team leader asks you to assess how this might affect competition in the market.

The demand for groceries is given by $Q = 1 - P$, and the firms compete à la Cournot. The firms' marginal costs, c_A and c_B, are determined by their respective purchase prices.

a. Initially, the two producers have identical purchase prices: $c_A = c_B = 0.25$. What is the Nash equilibrium in this case? Calculate and illustrate in a figure showing the reaction functions.

b. Through negotiations with suppliers, firm A manages to obtain a lower purchase price, so that $c_A^{low} = 0.2$, while B must pay a higher price, $c_B^{high} = 0.6$ (you can think of it as the suppliers giving with one hand and taking with the other to avoid losing money). What is the new Nash equilibrium? Calculate and illustrate in a figure showing the reaction functions.

c. Discuss how total output and the market price are affected by the suppliers' discriminatory pricing between firms A and B.

12.4   Advertising, Customer Loyalty and Profits

Alpha Lab and Beta Solutions are Bertrand duopolists. Their marginal costs are identical, and for simplicity we assume they are zero: $c_A = c_B = 0$.

a. The demand for Alpha's and Beta's products is given by $Q_A = 1 - 2p_A + p_B$ and $Q_B = 1 - 2p_B + p_A$, respectively. What is the Bertrand equilibrium in this case? What are the profits (which here are the same as revenue since costs are zero)? Calculate and illustrate.

b. Alpha launches an advertising campaign that increases customer loyalty. Assume that after the campaign, demand becomes $Q_A' = 1 - 1.5p_A + p_B$, that is, less sensitive to price increases for product A. The demand for Beta's product remains $Q_B = 1 - 2p_B + p_A$.

What will the price of the two products be in this case, and what are the firms' profits? Is Alpha's advertising campaign bad news for Beta? Calculate and illustrate.

12.5   Bertrand Leader

Let's revisit the Bertrand competitors Alpha Lab and Beta Solutions. The two firms are fully symmetric (we ignore advertising here), with the demand for Alpha's product being $Q_A = 1 - 2p_A + p_B$ while the demand for Beta's product is $Q_B = 1 - 2p_B + p_A$, and zero marginal cost $c_A = c_B = 0$.

Assume that Alpha is the leader and Beta is the follower—that is, Alpha sets its price first.

a. Compared to the simultaneous-move (Bertrand) case, will Alpha—as the first mover—set a higher or lower price?
b. Does Beta lose out because Alpha moves first?

12.6   Tips

12.1 Is a Merger Profitable?

Math Boxes 12.1 (duopoly) and 11.1 (monopoly) are useful for solving this problem, which allows you to practice the Cournot model for oligopoly. The application here is a merger, which we interpret as two firms joining to become one.

In part (a), there are duopolists who form a monopoly. This should be standard: you can simply compare the duopoly solution with the monopoly solution, both of which are thoroughly covered in the textbook.

Part (b), however, is a bit different, since it starts with a market of three producers— a triopoly. With the merger between two of them, the market structure becomes a duopoly. Although we have not gone through the triopoly case in the textbook, it should not be too difficult to find the market equilibrium since we assume the three firms are identical. After setting up the profit function for one firm and maximizing with respect to quantity, you can find the equilibrium quantity in the triopoly by setting $Q_A = Q_B = Q_C$ in the first-order condition, which must hold due to symmetry. From this, you can calculate the equilibrium price and profits.

12.2 Foreign Ownership, Competition and Social Welfare

Again, math boxes 12.1 (duopoly) and 11.1 (monopoly) are useful. Usually, we think increased competition is beneficial for the economy, but what if the competitor is foreign owned, so that its producer surplus leaves the country? The home country's economic surplus then consists of consumer surplus and the producer surplus of the *home* producer.

This problem is not technically very demanding but provides interesting insight into the pros and cons of foreign entry in a market with imperfect competition.

12.3 Lower Costs, More Expensive Food?

The goal of this exercise is to give you practice in analysing the Cournot model where the duopolists have different marginal costs. The Competition Authority is concerned that a dominant grocery chain has excessively favourable purchasing conditions, which could reduce competition and ultimately harm consumers.

You will find Math Box 12.1 useful in this exercise, as it shows how quantities and prices in the Nash equilibrium are affected by changes in marginal costs.

12.4 Advertising, Customer Loyalty and Profits

Math Box 12.3 is helpful for solving this problem, particularly the answer to part (a) follows quite directly from it. In part (b), we introduce an asymmetry in demand caused by Alpha's advertising campaign. This changes Alpha's reaction function and thus leads to a new Bertrand equilibrium. Calculate the implications for the profits of the two firms based on the new equilibrium prices. Pay special attention to what happens to Beta's price and profits following the advertising for product A!

12.5 Bertrand Leader

What happens if we have a leader–follower structure in a duopoly with price competition? This problem familiarises you with the Bertrand model and provides interesting insights, also highlighting the contrast between price and quantity competition. Math Box 12.3 is useful here but also see Math Box 12.2 for how to find equilibrium in a leader–follower situation (though that example uses quantity as the strategic variable).

What happens to the leader's price compared to the simultaneous-move case? And what about the follower's price? In the model with quantity competition (the Stackelberg model), the leader benefits at the expense of the follower: does the same hold here?

Game Theory

13

Anna and Brian on Their First Date

This chapter introduces game theory, which can be used to study decision making in a wide range of contexts. We go through two main categories of models: games with simultaneous moves and games with sequential moves. The exercises in this chapter give you practice in using game theory across a variety of exciting applications, from cooperation problems in study groups to global climate policy.

13.1   Two Games, abcd

a. Consider the game below, based on Table 13.1 and 13.2 in the textbook. Assume that the players achieve the highest total utility if both choose the high price.

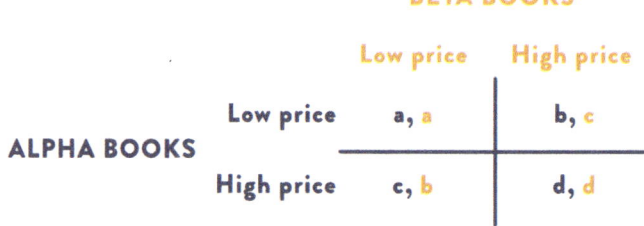

BETA BOOKS

		Low price	High price
ALPHA BOOKS	Low price	a, a	b, c
	High price	c, b	d, d

What must the ranking of a, b, c and d be for this to be a Prisoner's Dilemma game?

b. Consider next the game below, based on Table 13.3 in the textbook. What ranking of a, b, c and d makes this a Stag Hunt game?

EAST

		Brown	Green
WEST	Brown	a, a	b, c
	Green	c, b	d, d

c. Let's continue with the Brown vs. Green technology choice described in the table above. In Table 13.3 of the textbook, $a = 1$ and $b = 2$. But what if we swap the values of a and b so that $a = 2$ and $b = 1$ (while keeping $c = 0$ and $d = 3$)? Does this make any difference to the outcome? What do you think is the most logical ranking of a and b in this game, which is about the green transition?

13.2 Back in the Office After the Pandemic

You manage a firm that employs two workers, Audrey and Beth. During the pandemic, the firm adopted a work-from-home policy. Now that the pandemic is over, employees are free to choose whether to work from home or at the office. The table below shows each worker's utility from the two options.

BETH

		Home	Office
	Home	1, 1	3, 0
AUDREY			
	Office	0, 3	2, 2

a. How would you explain the payoffs in the matrix? What is the Nash equilibrium of this game?

As a manager, you believe in physical presence and would like to bring your employees back to the office. You don't want to force them back, as this may lead to tensions you wish to avoid. Accordingly, you decide to offer free lunch at work, making it more attractive to show up at the office voluntarily. With the new lunch policy, your workers have the utility shown in the table below:

BETH

		Home	Office
	Home	1, 1	2, 0
AUDREY			
	Office	0, 2	3, 3

b. Again, give your explanation of the payoffs in the matrix. What is the Nash equilibrium (or equilibria) of this game? Will the free lunch necessarily get your workers back to the office?
c. You make a phone call to Audrey, telling her about the free lunch and asking her to decide if she wants to come to the office (but leaving the choice entirely up to her). You tell her: "Once you've made your mind up, I will call Beth and inform her about your choice." Will this make a difference relative to your findings in part (b)?

13.3   Anna and Brian on Their First Date

Brian has long had a crush on Anna (to be honest, that's why he came to The Mill in the first place), and he has now gathered the courage to ask if they should do something together one evening. The question is what. It's a small town, and nightlife options are limited: the choice is between the cinema or the café.

It's their first date, and both are a bit nervous about suggesting something the other might not like. So, they decide to each write down on a slip of paper where they want to go—either cinema or café—and then show the slips to see if it works out.

Brian prefers the cinema, while Anna prefers the café. But above all, they want to spend the evening together. The alternative is that one goes to the cinema and the other to the café, which would make for a rather bad first date!

Brian wonders whether he should write "cinema" or "café" on his slip. He knows Anna prefers the café, so if he writes "café", Anna will be happy—but what if she thinks the same way and wants to make Brian happy, and she therefore writes "cinema"?

What is the equilibrium, or equilibria, in such a game? Set up the game matrix and find the equilibrium if meeting at the cinema gives Brian a payoff of 2 and Anna 1, meeting at the café gives Anna a payoff of 2 and Brian 1 and not meeting gives both of them a payoff of zero.

13.4   More Diet Books?

New celebrity chefs have stepped forward, making it possible for publishers to release a diet book twice a year, in both January *and* May.

a. Starting from Table 13.4 in the textbook, assume that the profits from releasing two books a year is the sum of the profits from releasing books in January and May. Add a row and a column for releasing a book in both January and May, so that you get a matrix with nine cells. What is the Nash equilibrium (or Nash equilibria)?
b. Now assume that Alpha Books moves first, for example by announcing in the press and their website when the new diet book(s) will be released. Does this give Alpha Books a first-mover advantage?

13.5   Brian and His Mother Again

This final exercise is about Brian's first year at the School of Economics. You may recall he faced some challenges back then, including financial ones. His mother, Audrey, would like to help him with money for food, but she also thinks it's high time he got a job. Brian loves food (especially frozen pizzas), but he doesn't like working.

Audrey's choice is whether to help her son or not, while Brian's choice is whether to work or not. The table below shows the utilities for mother and son in the different outcomes, where the first number in each cell is Brian's utility and the second is Audrey's utility.

AUDREY

		Help? No	Help? Yes
	Work? No	1, 0	3, 0.25
BRIAN			
	Work? Yes	0, 2	2, 3

Assume Audrey and Brian play a sequential game: Brian first chooses whether or not to work, and then Audrey decides whether or not to help him.

a. Audrey tells her son: "If you don't work, I won't give you anything!" Is this a credible threat?
b. Audrey decides to use up all her savings, so to support her son, she would have to take out a high-interest consumer loan. This imposes an additional cost of 0.5 on offering help. She repeats: "If you don't work, I won't give you anything!" Is her threat credible now?

13.6   Tips

13.1 Two Games, abcd

The goal of exercise a is to make you more familiar with the logic behind the Prisoner's Dilemma. Start by understanding what it takes for one player to be tempted to deviate from the joint best outcome (here, high prices). What does this imply about the relative sizes of the payoffs? Then think: what must the payoff be for the other player to respond in kind, that is, to do the same as the competitor?

Part (b) is a Stag Hunt, with multiple equilibria, but where one is preferred by both players, in this case the green transition. What must be true for this to be an equilibrium, meaning no player wants to deviate from it? At the same time, there should also be a "bad" equilibrium in this game, where both choose brown technology.

In part (c) you should look at the relative size of a and b in the payoff matrix, and discuss what you think is more reasonable, a > b or b > a, in the current context, which is a game about implementing green technology. There is no right or wrong answer here but use economic intuition to judge what makes most sense.

13.2 Back in the Office After the Pandemic

The first game, without any free lunch, is a classical Prisoner's Dilemma. Try to think of what could lie behind the utilities of Audrey and Beth in this case: there could be several stories here, so just use your imagination (and common sense).

When you as the leader of the firm introduce a free lunch, the game changes into a coordination game, with multiple equilibria. Again, use your own words to tell a story about Audrey and Beth that makes sense.

When you in part (c) place Audrey in a first-mover position, use the game in the extensive form and use backward induction to find what will be the equilibrium in this game.

Technically, I don't think this exercise is too challenging, but it presents you with a nice illustration of how game theory can be used to analyse situations facing many workplaces after the pandemic.

13.3 Anna and Brian on Their First Date

With the information given in the exercise, the game looks as follows:

Use the tick-off method to find the equilibrium or equilibria. Which of the games covered in the main textbook does this remind you of?

13.4 More Diet Books?

To understand what the profits are when we allow the publishers to release two books each year, look at the situation where Beta Books only releases a book in January, while Alpha Books releases books both in January and May—that is, the bottom-left cell. Here, Beta Books earns a profit of 1.5 from its January sales (since Alpha Books also releases a book then), while Alpha Books earns a profit of 1.5 from January plus 2 from May, totalling 3.5.

In part (b) you should use the tick-off method to find the equilibrium (or equilibria) and then discuss what you think would happen if one of the publishers could move before the other.

13.5 Brian and His Mother Again

This exercise gives you practice in analysing credible and non-credible threats. It also provides an important insight: that it can be advantageous to "tie yourself to the mast", meaning to limit your own available actions. The application is the relationship between Brian and his mother, where the mother wants Brian to work but often ends up helping him if he doesn't earn his own money. And this obviously does not exactly encourage Brian's willingness to work!

Use the payoff matrix to set up a game tree with Brian as the first mover: he chooses whether to work or not, and Audrey responds by either helping him out or not. The equilibrium of this game can be found in the usual way, by sawing off branches as shown in Chapter 13.5 of the textbook.

In part (b), Audrey puts herself in a position where she must take out expensive loans if she is to help her son. This affects Audrey's utility from supporting him, so you must reduce her relevant payoffs accordingly at the bottom of the game tree: How does this affect the equilibrium? As you will see, it can be an advantage for Audrey to limit her own choices!

Solutions

K. Bjorvatn, *Workbook for Microeconomics Made Simple*, Classroom Companion:
Economics, https://doi.org/10.1007/978-3-032-06357-1

Chapter 1 Income, Prices and Preferences

1.1 Higher Electricity Prices and Shifts in the Budget Line

a. The budget line for $I = 1$ and $p_A = p_B = 1$ is shown as m_1 in Fig. 1.1, with the maximum consumption of each good equal to 1 (as shown by the intercepts on the axes). Since Anna has decided to spend the same amount of money on both goods, she chooses $A^a_{Anna} = B^a_{Anna} = 0.5$, as illustrated at point a.

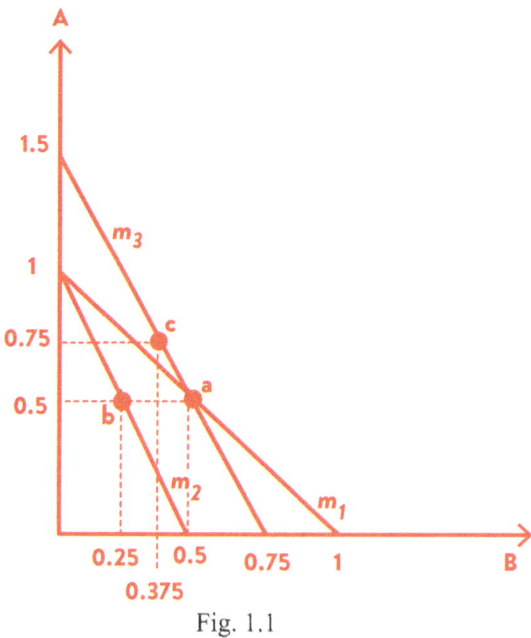

Fig. 1.1

b. When the price of good B increases, the budget line becomes steeper and rotates from m_1 to m_2. With the new higher price $p^h_B = 2$, the slope of the budget line becomes $-\frac{\partial A}{\partial B} = \frac{p_B}{p_A} = 2$.

The maximum consumption of good B is now 0.5, while the maximum consumption of A remains unchanged at 1. Since Anna continues to spend equal amounts on each good, her consumption of B falls accordingly.

Anna's consumption of B drops to 0.25, while the consumption of A remains unchanged at $A = 0.5$, that is, $A^b_{Anna} = 0.5$, $B^b_{Anna} = 0.25$, as illustrated at point b.

c. The government offers a subsidy per unit consumed of $s_B = 1$, so that the price the consumer pays returns to its original level. Since Anna consumes $B = 0.5$ at the lower price $p^l_B = 1$, the cost of the electricity subsidy for the government is $S_B = s_B B = (1)0.5 = 0.5$.

d. A cash transfer $S = S_B = 0.5$ shifts the budget line outward, shown as m_3 in Fig. 1.1, with the same steep slope as under the higher electricity price. Since the cash transfer is equal in value to the price subsidy, m_3 must pass through point a:

the cash transfer makes it possible to buy exactly as much electricity as with the price subsidy.

However, since Anna has decided to spend the same amount of money on each of the two goods, she will not choose point a, but rather point c, where $A = 0.75$, and $B = 0.375$. We can verify that this consumption bundle satisfies the budget constraint:

$$I + S = p_A A + p_B^h B \Rightarrow 1.5 = 1(0.75) + 2(0.375)$$

1.2 Anna and Her Friend Deal with Higher Electricity Prices

a. We know from Math Box 1.2 that maximising a Cobb-Douglas utility function $U = A^\alpha B^{1-\alpha}$ gives:

$$A = \frac{\alpha I}{p_A} \Rightarrow p_A A = \alpha I$$

Similarly, for good B we have:

$$B = \frac{(1 - \alpha)I}{p_B} \Rightarrow p_B B = (1 - \alpha)I$$

Since Anna spends equal amounts on each good, we have:

$$p_A A = p_B B \Rightarrow \alpha I = (1 - \alpha)I \Rightarrow \alpha_A = 0.5$$

Her utility function must therefore be $U_{Anna} = A^{0.5} B^{0.5}$.

Bella spends three times as much on her home base as on activities, which implies that:

$$3p_A A = p_B B \Rightarrow 3\alpha I = (1 - \alpha)I \Rightarrow \alpha_B = 0.25$$

Her utility function is therefore $U_{Bella} = A^{0.25} B^{0.75}$.

b. Before the increase in the price of electricity, we have $p_A = p_B = 1$. With income $I = 1$, and the utility weights that we found in part (a) above, Anna will therefore choose consumption in point a in Fig. 1.2a, with consumption $A_{Anna}^a = B_{Anna}^a = 0.5$, while Bella chooses point b with consumption $A_{Bella}^b = 0.25, B_{Bella}^b = 0.75$.

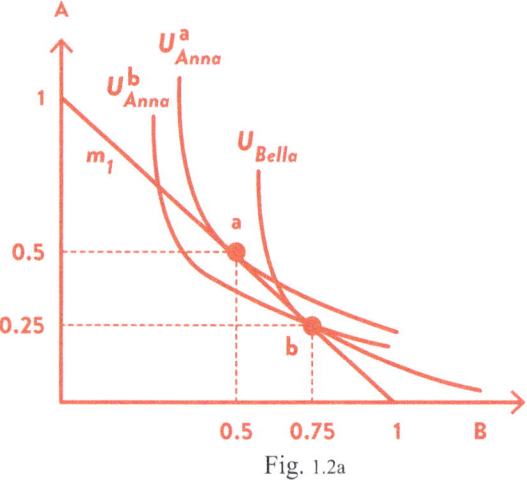

Fig. 1.2a

c. From Math Box 1.1 we know that with $\alpha = 0.5$, the marginal rate of substitution is given by $MRS = A/B$, so that with $A = 0.25, B = 0.75$ in point b, Anna would have:

$$MRS = \frac{A}{B} = \frac{0.25}{0.75} = \frac{1}{3}$$

This means that Anna at point b would be willing to give up only one third of A for a marginal increase the consumption of B, while the slope of the budget line shows that she needs to give up *one unit* of A to do so. Clearly, she's not willing to do that: on the contrary, $MRS < p_B/p_A$ implies that she would like to have more of A, and not more of B.

Her optimal consumption combination is at point a, where her MRS is identical to the slope of the budget line: at this point $A = B$ and here her MRS is perfectly aligned with the slope of the budget line.

At point b her utility is $U_{Anna}^b = (0.25)^{0.5}(0.75)^{0.5} \approx 0.43$, and at point a her utility is $U_{Anna}^a = (0.5)^{0.5}(0.5)^{0.5} = 0.5$.

d. When the price of electricity increases to $p_B^h = 2$, Anna will choose point c and Bella point d on the new budget line is m_2, as shown in Fig. 1.2b. We see from the expressions on optimal consumption in part (a) that their consumption of B has now been cut in half: $B_{Anna}^c = 0.25, B_{Bella}^d = 0.375$.

Since Bella was already spending more on housing, she will have to turn down the heating (i.e., reduce B) more than Anna will. We also note that the increase in electricity prices does not affect their demand for food. That's how it is with Cobb-Douglas preferences—the cross-price effect is zero.

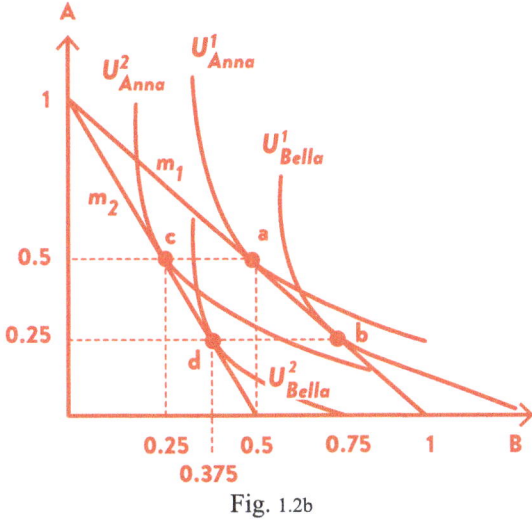

Fig. 1.2b

1.3 Price Support or Cash Transfer?

With a price subsidy that shields Anna from the increase in the market price of electricity, she chooses consumption at point a in Fig. 1.3—the same point as in Fig. 1.2b, with consumption $A_a = B_a = 0.5$. This gives her utility $U_a = A^{0.5}B^{0.5} = (0.5)^{0.5}(0.5)^{0.5} = 0.5$.

The size of the price subsidy is $S = 0.5$, as shown in Exercise 1.1. You argue that this amount should instead be given as a cash transfer. Anna's total income then becomes $I + S = 1.5$, and the budget line shifts outward in parallel (relative to the situation with a high electricity price but no support, line m_2) to a new line m_3, as shown in Fig. 1.3. This line passes through point a, since the cash transfer allows her to purchase exactly as much electricity as she could with the price subsidy.

However, with a cash transfer, Anna will not choose point a. With a balanced Cobb-Douglas utility function, we know that Anna wants to spend equal amounts of money on each good (just as in Exercise 1.1, although there we simply assumed it).

With total income after the cash transfer $I + S = 1.5$, and prices $p_A = 1$ and $p_B = 2$, we then get $A = 0.75$ and $B = 0.375$, as shown as point c in Figure 1.3.

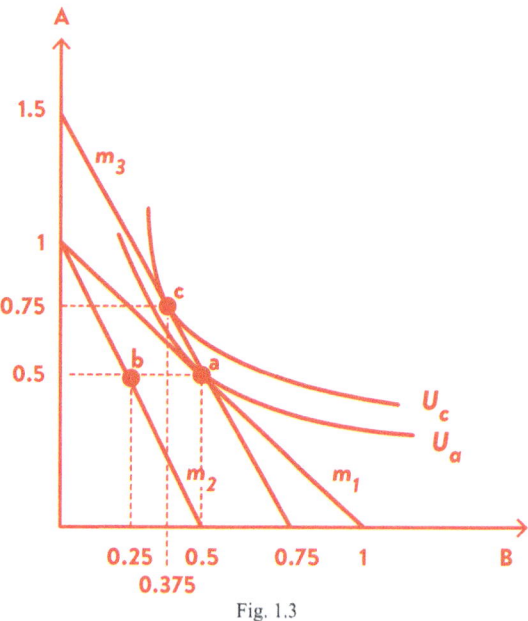

Fig. 1.3

Her utility at point c is:

$$U_c = A^{0.5}B^{0.5} = (0.75)^{0.5}(0.375)^{0.5} \approx 0.53$$

We see that $U_c > U_a$: a cash transfer gives Anna higher utility than an equivalent amount in electricity price subsidy. The reason is that a cash transfer gives Anna greater freedom to allocate her money as she wishes, while a price subsidy is tied to the consumption of one specific good.

We also see that Anna consumes less electricity (lower B) with the cash transfer than with the subsidised electricity price. A cash transfer is therefore a better policy for supporting consumers.

1.4 Peanuts + Beer = True Love

a. With this type of preferences (perfect complements), the indifference curves are L-shaped: the goods go together, like left and right shoes, and having one extra of either shoe gives you no additional satisfaction. Utility is defined by the minimum term in the utility function (for example, if you have 2 left shoes and 3 right shoes, it's the number of left shoes that determines your utility from the pair). This means that the consumer will choose A and B such that $\frac{A}{\alpha} = \frac{B}{1-\alpha}$.

No other combination gives higher utility and will therefore not be chosen. To see why, consider a situation where $\frac{A}{\alpha} > \frac{B}{1-\alpha}$. In this case, utility is defined by the smaller term, $\frac{B}{1-\alpha}$, and one can safely reduce A without lowering utility (like giving away an extra left shoe—you still only have one usable pair, and that extra shoe didn't make any differences anyway).

The optimal consumption combination is thus $\frac{A}{B} = \frac{\alpha}{1-\alpha}$, which can be written as $A = \frac{\alpha}{1-\alpha}B$. We now plug this into the budget constraint $I = p_A A + p_B B$ and get

$$I = p_A \left(\frac{\alpha}{1-\alpha} \right) B + p_B B$$

$$B = \frac{(1-\alpha)I}{\alpha p_A + (1-\alpha)p_B} \quad \text{Optimal consumption of B}$$

We then plug this value back into the expression for A:

$$A = \left(\frac{\alpha}{1-\alpha} \right) B = \left(\frac{\alpha}{1-\alpha} \right) \frac{(1-\alpha)I}{\alpha p_A + (1-\alpha)p_B}$$

$$= \frac{\alpha I}{\alpha p_A + (1-\alpha)p_B} \quad \text{Optimal consumption of A}$$

b. We are told in the exercise that Brian has a strong preference for beer, with $\alpha_B = \frac{1}{3}$, while Anna likes peanuts more, with $\alpha_A = \frac{2}{3}$. With income $I = 1$ and initial prices $p_A = p_B = 1$, we can use the expressions for optimal consumption derived in part (a) to find Brian's consumption as (with superscript "l" indicating that this is the scenario where the price of beer is low):

$$A^l_{Brian} = \frac{\frac{1}{3}}{\frac{1}{3}(1) + \frac{2}{3}(1)} = \frac{1}{3}$$

$$B^l_{Brian} = \frac{\frac{2}{3}}{\frac{1}{3}(1) + \frac{2}{3}(1)} = \frac{2}{3}$$

This is marked as point a in Fig. 1.4a. For Anna we have:

$$A^l_{Anna} = \frac{\frac{2}{3}}{\frac{2}{3}(1) + \frac{1}{3}(1)} = \frac{2}{3}$$

$$B^l_{Anna} = \frac{\frac{1}{3}}{\frac{2}{3}(1) + \frac{1}{3}(1)} = \frac{1}{3}$$

This is marked as point b in Fig. 1.4a.

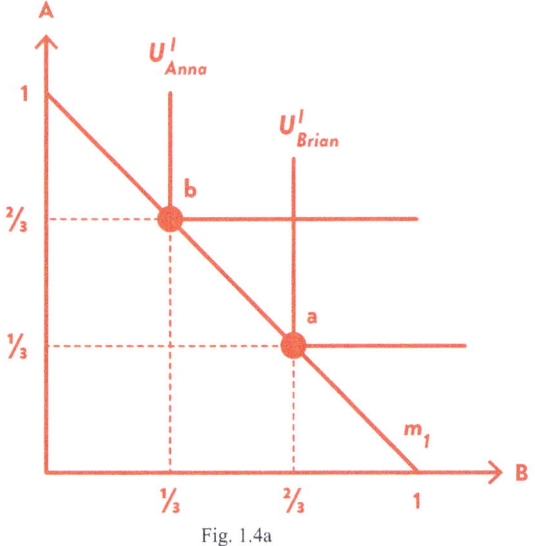

Fig. 1.4a

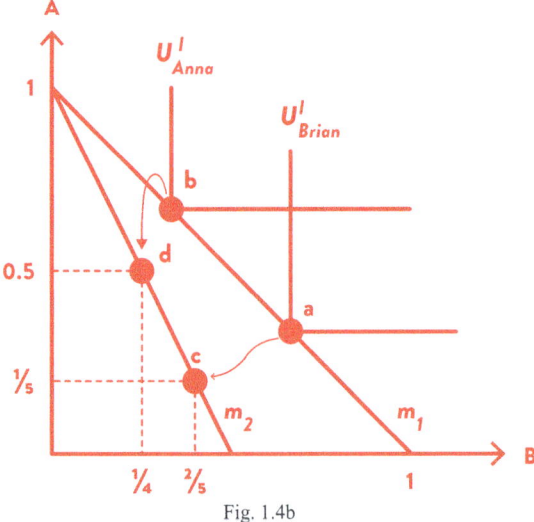

Fig. 1.4b

c. When the price of beer increases to $p_B^h = 2$ (superscript "h" indicating the *high* price), we get:

$$A_{Brian}^h = \frac{\frac{1}{3}}{\frac{1}{3}(1) + \frac{2}{3}(2)} = \frac{3}{15} = \frac{1}{5}$$

$$B_{Brian}^h = \frac{\frac{2}{3}}{\frac{1}{3}(1) + \frac{2}{3}(2)} = \frac{6}{15} = \frac{2}{5}$$

In Fig. 1.4b this is shown as point c (to make the figure clearer I have not included the new indifference curves). For Anna we then have:

$$A_{Anna}^h = \frac{\frac{2}{3}}{\frac{2}{3}(1) + \frac{1}{3}(2)} = \frac{1}{2}$$

$$B_{Anna}^h = \frac{\frac{1}{3}}{\frac{2}{3}(1) + \frac{1}{3}(2)} = \frac{1}{4}$$

This corresponds to point d in Fig. 1.4b. For Brian, we see that his consumption of beer goes down with $B_B^l - B_B^h = \frac{2}{3} - \frac{2}{5} = \frac{4}{15}$. In percentage terms, Brian's consumption of beer is reduced by:

$$\frac{B_{Brian}^l - B_{Brian}^h}{B_{Brian}^l} = \frac{\frac{4}{15}}{\frac{2}{3}} = \frac{2}{5} = 40\%$$

For Anna, the reduction is $B_{Anna}^l - B_{Anna}^h = \frac{1}{3} - \frac{1}{4} = \frac{1}{12}$, which in percentage terms equals:

$$\frac{B_{Anna}^l - B_{Anna}^h}{B_{Anna}^l} = \frac{\frac{1}{12}}{\frac{1}{3}} = \frac{1}{4} = 25\%$$

The reduction in beer consumption due to the price increase is therefore greatest—both in absolute and relative terms—for the person who initially consumed the most beer, namely Brian.

1.5 How Does an Interest Rate Increase Affect the Demand for Housing and Other Goods?

From Math Box 1.2 we see that with $\alpha = 0.5$ (a balanced CD utility function) the optimal consumption is given by:

$$A = 0.5\frac{I}{p_A} \quad \text{Optimal consumption of A, balanced CD}$$

$$B = 0.5\frac{I}{p_B} \quad \text{Optimal consumption of B, balanced CD}$$

What about the case where housing is a basic good, that is, $U = A^{0.5}(B - b)^{0.5}$? We can apply the same method as for a standard utility function, but with replacing B with $B - b$. With $\alpha = 0.5$, the marginal rate of substitution is now (see Math Box 1.1):

$$MRS = \frac{A}{B - b}$$

In a consumption optimum, MRS must equal the price ratio (see Math Box 1.2):

$$\frac{A}{B-b} = \frac{p_B}{p_A}$$

This can be expressed as:

$$A = (B-b)\frac{p_B}{p_A}$$

We plug this expression into the budget constraint:

$$I = p_A(B-b)\frac{p_B}{p_A} + p_B B$$

$$I = 2p_B B - p_B b$$

From this we can find the consumption of B as:

$$B = \frac{I + p_B b}{2p_B} \quad \text{Optimal consumption of B, when B is a basic good}$$

We use this expression for B and substitute it into the budget constraint, yielding the optimal consumption of A as:

$$I = p_A A + p_B \left(\frac{I + p_B b}{2p_B}\right)$$

$$I = p_A A + \left(\frac{I + p_B b}{2}\right)$$

$$A = \frac{I - p_B b}{2p_A} \quad \text{Optimal consumption of A, when B is a basic good}$$

We then use the numbers from the exercise and illustrate the effect of the interest rate increase using the two different utility functions, as shown in Figs. 1.5a and 1.5b. Initially, the consumer has chosen point a, which, with balanced Cobb-Douglas utility, means $A = B = 0.5$. When B is a basic good, the starting point is $A = 0.4, B = 0.6$.

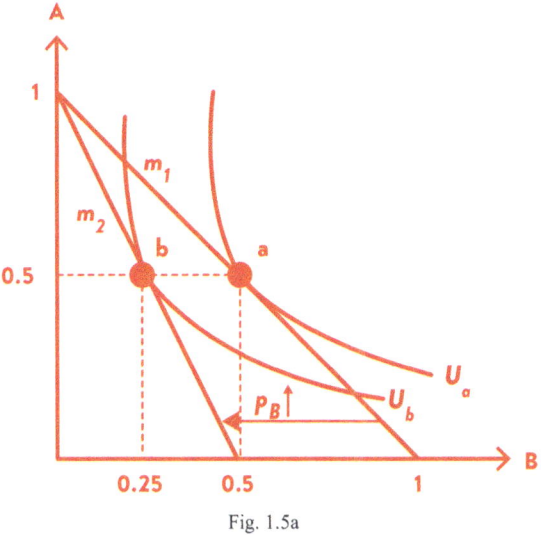

Fig. 1.5a

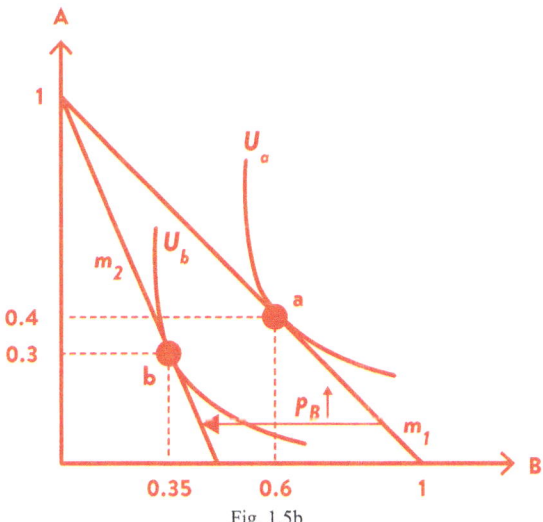

Fig. 1.5b

The price increase leads to a movement from point a to point b. With Cobb-Douglas utility, this means a reduction in the consumption of B by 0.25, and no change in the demand for A (the cross-price effect is zero). However, the picture is slightly different when B is a basic good. In absolute terms, the demand for housing falls just as much as with Cobb-Douglas utility—a reduction of 0.25 (from 0.6 to 0.35). But here, the cross-price effect is negative: the demand for A falls from 0.4 to 0.3.

Intuitively, the consumer seeks to limit the negative effect of the price increase on the basic good, housing, and therefore also reduces demand for other goods.

Chapter 2 More About Consumer Choice

2.1 The Effect of Higher Electricity Prices, in Two Steps

Anna has $\alpha = 0.5$, while Bella has $\alpha = 0.25$. From Math Box 2.1, we know that the price increase of good B from the low price, $p_B^l = 1$ to the high price $p_B^h = 2$, leads to a reduction in Anna's consumption of good B from 0.5 to 0.25. As shown in the Math Box, the substitution effect is a reduction from 0.35 to 0.25.

$$B_{Anna}^a(0.5) \rightarrow B_{Anna}^b(0.35) \quad SE_{Anna}^B \approx -0.15$$

$$B_{Anna}^b(0.35) \rightarrow B_{Anna}^c(0.25) \quad IE_{Anna}^B \approx -0.1$$

This is illustrated in Fig. 2.1a.

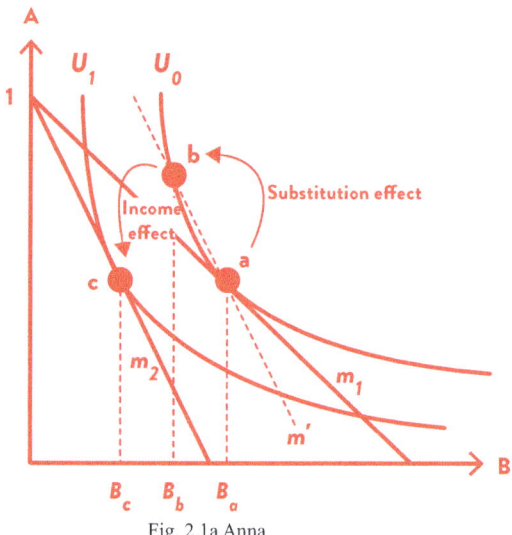

Fig. 2.1a Anna

But what about Bella? From Math Box 1.2 we see that:

$$\frac{A}{B} = \frac{\alpha p_B}{(1-\alpha)p_A} \quad \text{Optimal consumption combination}$$

$$B = \frac{(1-\alpha)I}{p_B} \quad \text{Optimal consumption of B}$$

$$A = \frac{\alpha I}{p_A} \quad \text{Optimal consumption of A}$$

Bella has $\alpha = 0.25$ and with an income $I = 1$ and with the low electricity $p_B^l = 1$, she will consume:

$$B_{Bella}^a = \frac{(1-\alpha)I}{p_B} = 0.75$$

$$A_{Bella}^a = \frac{\alpha I}{p_A} = 0.25$$

We now plug Bella's optimal consumption into her utility function and follow the same procedures as in Math Box 2.1. Step 1 is to find the initial level of utility:

$$U_{Bella}^a = \left(A_{Bella}^a\right)^{0.25}\left(B_{Bella}^a\right)^{0.75} = 0.25^{0.25}0.75^{0.75} \approx 0.57$$

Step 2 is to use the optimal consumption bundle and find the consumption of B at the new price level. When the price of B doubles to $p_B^h = 2$, we get:

$$\frac{A}{B} = \frac{\alpha p_B}{(1-\alpha)p_A} = \frac{0.25(2)}{0.75(1)} = \frac{2}{3}$$

This means that $A = \frac{2}{3}B$. Substitute this into Bella's utility function, and we get:

$$U_{Bella}^b = \left(\frac{2}{3}B\right)^{0.25}B^{0.75} = B\left(\frac{2}{3}\right)^{0.25}$$

Since the substitution effect is a movement along the original indifference curve, with utility approximately equal to 0.57, we can use the above expression and find Bella's consumption of B at point b as:

$$B_{Bella}^b \approx 0.63$$

The substitution effect for Bella is therefore:

$$B_{Bella}^a(0.75) \rightarrow B_{Bella}^b(0.63) \quad SE_{Bella}^B \approx -0.12$$

Bella chooses the following consumption of B with the new, higher price:

$$B_{Bella}^c = \frac{(1-\alpha)I}{p_B^h} = \frac{0.75}{2} = 0.375$$

The income effect is therefore:

$$B_{Bella}^b(0.63) \rightarrow B_{Bella}^c(0.375) \quad IE_{Bella}^B \approx -0.26$$

This is illustrated in Fig. 2.1b. We notice that the size of the income effect (in the figure, the movement from point b to c) for Bella is much larger than for

Anna. Intuitively, since housing constitutes a much larger share of Bella's initial consumption, she is more severely affected than Anna by the increase in the electricity price.

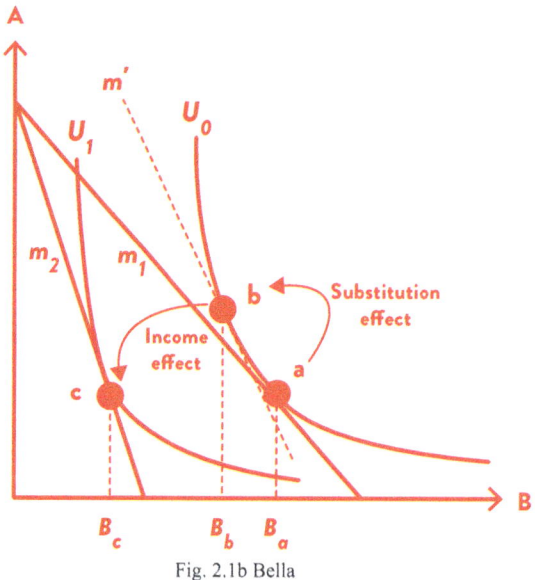

Fig. 2.1b Bella

2.2 Cash Support for Education in Mexico: Part I

a. The optimal consumption choice is given by the condition:

$$MRS \equiv \frac{MU_B}{MU_A} = \frac{p_B}{p_A}$$

that is, where the slope of the indifference curve, i.e., the marginal rate of substitution (MRS), equals the slope of the budget line, which is given by the price ratio between the two goods A and B. The Buenos family has a balanced Cobb-Douglas utility function, such that:

$$MU_A = \frac{\partial U}{\partial A} = 0.5A^{-0.5}B^{0.5} \quad \text{Marginal utility of A}$$

$$MU_B = \frac{\partial U}{\partial B} = 0.5A^{0.5}B^{-0.5} \quad \text{Marginal utility of B}$$

The slope of the indifference curve is then given by:

$$MRS = \frac{MU_B}{MU_A} = \frac{A}{B}$$

With prices $p_A = p_B = 1$, the price ratio and thus the slope of the budget line is equal to one. Therefore, at the optimum:

$$MRS = \frac{p_B}{p_A} \Rightarrow \frac{A}{B} = 1 \Rightarrow A = B$$

We use this information in the budget constraint $I = p_A A + p_B B$. Since the income without support is $I = 1$ and prices are $p_A = p_B = 1$, the budget constraint simplifies to $1 = A + B$ and since we know the family will optimally choose $A = B$, we get:

$A = B = 0.5$ Optimal consumption of A and B without support

This is indicated by point a in Fig. 2.2a.

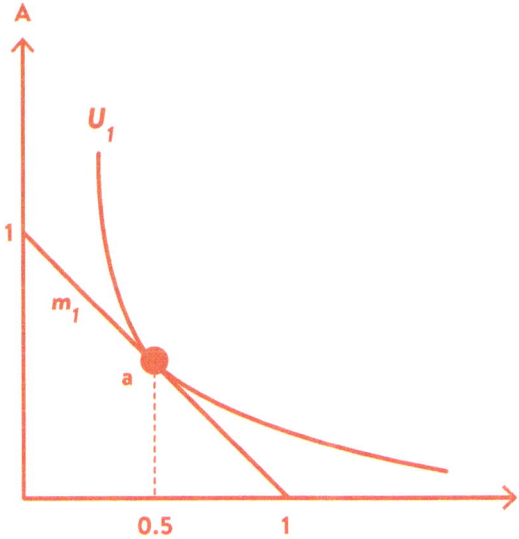

Fig. 2.2a Bueno's choice without support

b. With an unconditional cash support $S = 0.5$, the income is now $I' = I + S = 1 + 0.5 = 1.5$ and the budget constraint becomes $1.5 = A + B$.

Since the Buenos family chooses to consume equal amounts of the two goods, $A = B$, we now have:

$A = B = 0.75$ Optimal consumption of A and B with unconditional support

In Fig. 2.2b, we see that the cash support shifts the budget line from m_1 to m_2, resulting in a new equilibrium at point b.

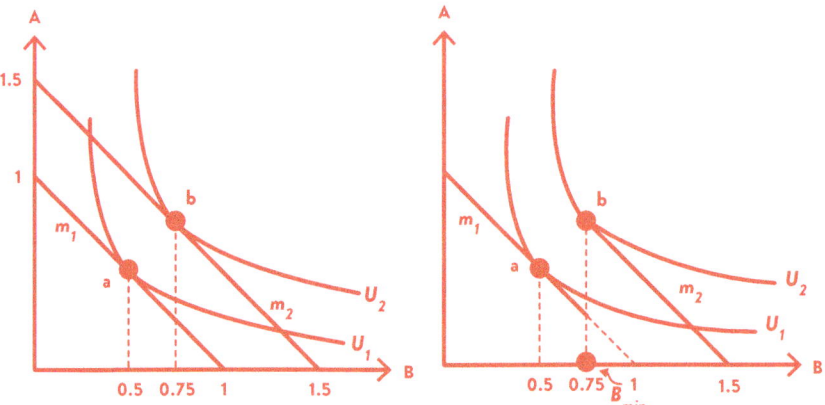

Fig. 2.2b Bueno's choice, unconditional support. Fig. 2.2c Bueno's choice, conditional support.

c. Fig. 2.2c illustrates the effect of conditional support, which here means that the family must meet a minimum requirement $B_{min} = 0.75$. Without support, the budget line is given by m_1, while with support S it is given by m_2. We see that when the support is conditional, there is a kink in the budget line: for $B \leq B_{min}$, the conditions are not met, and the family does not qualify for the support.

Since the Buenos family would have chosen $B = 0.75$ anyway, the condition is not binding for them: the family moves from point a to point b, just as with the unconditional cash support.

2.3 Cash Support for Education in Mexico: Part II

a. The Aires family has the same income as the Buenos family but places less weight on their children's education in their utility function $U = A^\alpha B^{1-\alpha}$, meaning $\alpha > 0.5$. Figure 2.3a illustrates. With the higher weight on good A relative to good B, they will choose a point higher up on the budget line m_1 without support, such as point a. An unconditional cash support shifts the budget line parallel upwards to m_2, and the new equilibrium will be at point c, where the new indifference curve U_3 is tangent to m_2.

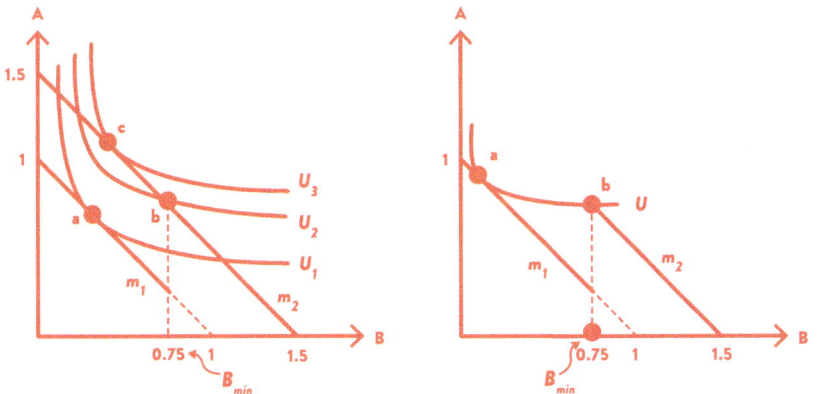

Fig. 2.3a Aires' choice   Fig. 2.3b Conditional support? No, gracias

What about conditional cash support? We see that point c lies to the left of the minimum requirement $B_{min} = 0.75$, so for the Aires family to qualify for the conditional support, they must spend more on education. If they accept the gift with the conditions, they will choose point b, which is the kink point on budget line m_2 (this corner solution gives them the highest utility, given that they must meet the support conditions). The Aires family must therefore spend more on schooling and less on other goods than they would with unconditional cash support. Their utility in this case is U_2, which is lower than U_3.

Note, however, that if α is sufficiently high, the family will refuse the conditional cash support. Figure 2.3b illustrates this. With these preferences, the Aires family is indifferent between accepting or rejecting the conditional gift (you can see this because the optimal choice on budget line m_1, at tangency point a, yields the same utility as in the corner solution with the gift, i.e., point b on m_2), and for values of α higher than this, they will prefer to decline it.

Can you find the critical level of α that makes the family indifferent between accepting the conditional support and rejecting it? We know from Math Box 1.2 in the textbook that $A = \alpha I / p_A$ and that $B = (1 - \alpha) I / p_B$, which with $p_A = p_B = I = 1$ implies $A = \alpha$ and $B = 1 - \alpha$, so that (as I showed you in the tip to this exercise) utility without support (at point a on the budget line m_1) is: $U_a = A^\alpha B^{1-\alpha} = (\alpha)^\alpha (1-\alpha)^{1-\alpha}$. This should be compared to the utility when the family receives support but has to fulfill the spending requirement on education, $B = 0.75$, implying that with support the family can also consume $A = 0.75$, given by point b on budget line m_2. Their utility in this case is $U_b = A^\alpha B^{1-\alpha} = (0.75)^\alpha (0.75)^{1-\alpha} = 0.75$. We can then show that $U_a = U_b$ for $\alpha \approx 0.92$. With an α higher than this, the Aires family would say "No, gracias" to the conditional support.

b. There is a major debate about which is better: conditional or unconditional support. Three common claims in this debate are:

A. It is necessary to make the transfer conditional to achieve the government's goals for schooling.

B. Making the support conditional on schooling makes it less valuable to families.
C. It makes no difference whether the support is conditional or unconditional.

For the Buenos family, claim C applies, while for the Aires family, claims A and B apply: unconditional support leads Aires to spend less on schooling than the authorities want, so conditional support is necessary to achieve the goal of having all children in school; and the Aires derives lower utility from conditional support than from unconditional support.

In summary, all the claims in the debate about unconditional versus conditional cash support can be valid, depending on the preferences of the households receiving the support.

2.4 Brian's Choice: Beer or Wine … or Both?
a. Brian's marginal utility of wine (A) is given by:

$$MU_A = \frac{\partial U}{\partial A} = 1$$

While his marginal utility of beer (B) is:

$$MU_B = \frac{\partial U}{\partial B} = \frac{1}{B}$$

As you can see, Brian's marginal utility of beer is extremely high when he initially has very little of it: MU_B approaches infinity as B approaches zero! This means that Brian will start by consuming beer before even considering buying wine. In the next part of the exercise, we'll find out exactly at what level of income he starts buying wine.
b. The marginal utility of beer decreases as he consumes more of it. Since the marginal utility of wine is constant, there will eventually come a point where he has enough beer in the fridge. If his income increases beyond this point, he will also start spending money on wine.

What level of income is required for him to start buying wine? Brian is indifferent between spending his last unit of income on beer or wine when $MU_B = MU_A$. As we saw in part (a), this gives:

$$MU_B = MU_A \Rightarrow \frac{1}{B} = 1 \Rightarrow B = 1$$

We also know that up until this point of indifference, he spends all his money on beer, since it gives him higher utility. We can think of $B = 1$ as Brian's *beer goal*—imagine it as a six-pack in the fridge. Up to $I = 1$, therefore, he spends all his money on beer. Any income beyond that he spends on wine.

Look at Fig. 2.4 which show the consumption of beer and wine as functions of income. Let's start at the origin: Brian has no money, and the fridge is empty. Then he starts earning money, and we move to the right along the horizontal income axis.

What happens to his beer consumption? It increases one-to-one with income along the B-line. And what happens to wine consumption? Nothing—at least not yet.

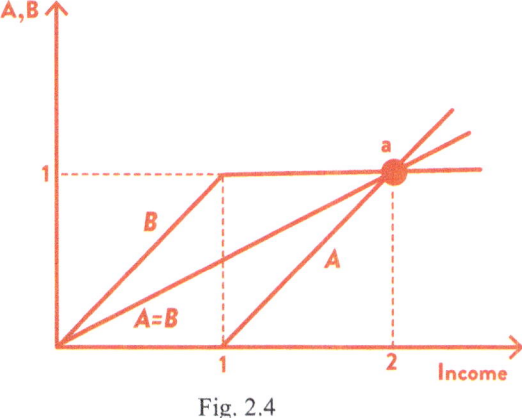

Fig. 2.4

But at some point, he reaches his beer goal, $B = 1$. And beyond that, all the money is spent on wine, along the A-line in the figure. At an income of $I = 2$ he will spend equal amounts on beer and wine, and for income levels beyond that, he spends more money on wine than on beer.

c. With Cobb-Douglas preferences, it's never the case that the consumer chooses only one of the goods (he would always have both beer and wine in the fridge, even with low income). And as income increases, consumption of both goods would increase—he would never spend all the extra money just on wine, for instance. In the case of balanced CD preferences, $U = A^{0.5}B^{0.5}$, and with equal prices, Brian would consume $A = B = 0.5I$, as shown by the $A = B$-line in the figure. So, the quasilinear utility function used in this exercise leads to somewhat different implications for consumption behaviour than a Cobb-Douglas utility function would.

d. Is B an inferior good in the case of quasilinear preferences? The answer is no. By definition, the consumption of an inferior good decreases when income increases (and increases when income falls). But as we see from the figure, with these preferences, beer consumption does not fall even when income rises beyond the critical level, it simply remains constant.

Of course, the share of beer in Brian's total beverage budget decreases as income rises beyond the critical threshold, and at some point, when income becomes sufficiently high, he will drink more wine than beer, but that's a different matter.

2.5 Conrad's and Anna's Saving

a. From Math Box 2.2, we know that optimal saving under a general Cobb-Douglas utility function is given by:

$$s = \alpha I_1 - (1 - \alpha)\left(\frac{I_2}{1 + r}\right)$$

Since both Conrad and Anna have $\alpha = 0.5$, this simplifies to:

$$s = 0.5\left(I_1 - \frac{I_2}{1 + r}\right)$$

We are told that Conrad has no income in period 2, so $I_2^C = 0$, and since $I_1^C = 1$ we see from the expression above that his saving becomes:

$$s_C = 0.5 \quad \text{Conrad's saving}$$

What about Anna? With $r = 0$ and $I_1^A = I_2^A$ we see from the expression above that Anna chooses not to save anything at all. Her choice is point a where $x_1^A = I_1^A = 0.5$.

$$s_A = 0 \quad \text{Anna's saving}$$

This makes intuitive sense: with a balanced utility function, equal income in both periods and a zero interest rate (which we can interpret as the price of consumption today being the same as tomorrow), she wants to consume the same amount in each period.

Figure 2.5a illustrates Conrad's budget constraint and optimal choice. With $I_1^C = 1$ and $r = 0$, the budget line is m_1, which intersects the axes at $x_1 = x_2 = 1$. We have seen that he chooses to save half his income, which with $I_1^C = 1$ implies $x_1^C = x_2^C = 0.5$, as illustrated at point a

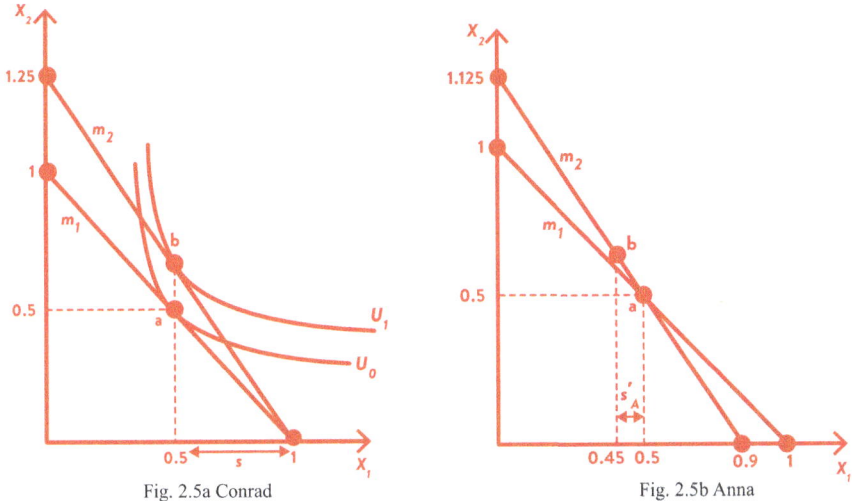

Fig. 2.5a Conrad   Fig. 2.5b Anna

Figure 2.5b shows Anna's budget constraint and her choice. With $r = 0$, the budget line is m_1. Her preferred consumption is at point a with equal consumption in both periods, and since her income is also the same in both periods, she chooses not to save.

b. With $r' = 0.25$, Conrad's budget line becomes m_2 in Fig. 2.5a. We see that it becomes steeper, rotating around the point where it intersects the horizontal axis (where Conrad's saving is zero and therefore unaffected by interest rate changes). The new budget line intersects the vertical axis at $x_2 = I_1(1 + r) + I_2$, which with the higher interest rate implies $x_2 = 1(1.25) + 0 = 1.25$.

As we saw in part (a), Conrad saves half of his income, $s_C = 0.5$, regardless of the interest rate. The increase in the interest rate therefore leads him to move from point a to point b, with the same consumption in period 1, and thus unchanged saving.

It may be useful to apply what we've learned about the substitution effect and the income effect to understand why saving does not change in Conrad's case. A higher interest rate means it becomes more expensive to consume today, and the substitution effect pushes towards lower consumption today and higher consumption tomorrow, that is, more saving.

But the income effect works in the opposite direction: Conrad is a saver, and with a higher interest rate, the value of his saving increases—in a sense, he becomes richer. He would like to use this increase in wealth to raise consumption in both periods, which pushes toward *less* saving (since more consumption today implies less saving).

We recognise from the analysis in Chapter 2 of the textbook that these two effects cancel out with Cobb-Douglas preferences.

What about Anna? If the interest rate rises to $r' = 0.25$, her budget line becomes m_2 in Fig. 2.5b. We see that it becomes steeper and rotates around the point where her saving is zero, that is, around point a. This is reasonable, since a change in the interest rate does not affect consumption possibilities when the consumer neither saves nor borrows.

The intercept of the budget line on the horizontal axis is given by $x_1 = I_1 + \frac{I_2}{1+r}$, which with the higher interest rate is $x_1 = 0.5 + \frac{0.5}{1.25} = 0.9$. The intercept on the vertical axis is $x_2 = I_1(1+r) + I_2$, which with the higher interest rate is $x_2 = 0.5(1.25) + 0.5 = 1.125$.

We know from the expression for optimal saving that with $r' = 0.25$, Anna will choose:

$$s = 0.5\left(I_1 - \frac{I_2}{1+r'}\right) = 0.5\left(0.5 - \frac{0.5}{1.25}\right) = 0.05 \equiv s'_A$$

The higher interest rate has therefore led to a move from point a to b in Fig. 2.5b (for simplicity of exposition, I have not included the indifference curves in this figure), implying that Anna has started saving.

c. We see that the interest rate increase leads to higher saving for Anna, but not for Conrad. How can this be explained? After all, they have the same preferences.

To answer this question, note how the interest rate increase affects Anna's budget line. It pivots around the point where saving is zero, shifting inward along the horizontal axis and outward along the vertical axis. This is equivalent to a combination of a lower price of consumption in period 2 (the new budget line intersects the vertical axis higher up) and a higher price of consumption in period 1 (it intersects the horizontal axis further to the left). And while a lower price of consumption in period 2 does not affect consumption in period 1 when preferences are Cobb-Douglas (the cross-price effect is zero, as is the case for Conrad), we know that a higher price of consumption in period 1 leads to lower consumption in period 1 (a negative own-price effect). It is this additional effect that makes the substitution effect stronger for Anna than for Conrad, and it must dominate the income effect (we know that, since Anna's saving always increases with a higher interest rate). This explains why Anna's saving increases, while Conrad's remains unchanged.

Intuitively, Conrad is a big saver, he must be, since he will have no income in his retirement (period 2). An increase in the interest is therefore necessarily good news for him: it gives him a significant positive income effect. And the positive income effect is an argument *against* increased saving. Anna, on the other hand, does not need to save so much, as she has income in both periods. The change in interest rate then does not create any big income effect, as her saving is small (in fact, she has zero savings to start out with), and for her, the story is mostly about a substitution effect, which is an argument in favour of increased saving.

Chapter 3 Consumers at Work

3.1 Audrey Likes to Work, Beth Only Works for the Money

Let's start with the simplest case—Beth. She has a standard, balanced Cobb-Douglas utility function, and as we know from Math Box 3.1, this means she will choose to divide her available time $T = 18$ equally between leisure and work:

$$F_B = 9$$

$$J_B = 9 \quad \text{Beth's choice of work hours}$$

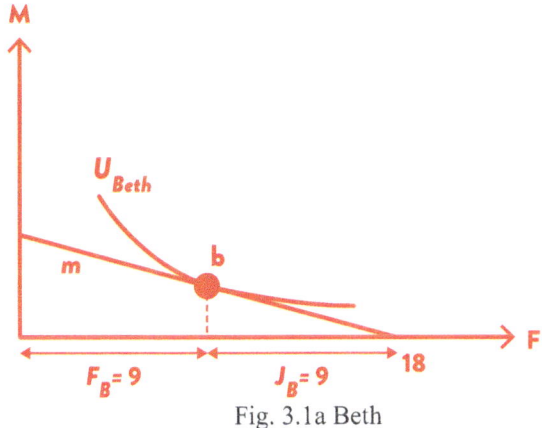

Fig. 3.1a Beth

Audrey's utility function is a bit different, as it includes three arguments: consumption, leisure and working hours.

$$U_A = (MFJ)^{\frac{1}{3}}$$

We follow the same approach as in Math Box 3.1. We substitute $J = T - F = 18 - F$ and can then write her utility function as:

$$U_A = (MF(18 - F))^{\frac{1}{3}}$$

We differentiate with respect to F (using the chain rule) to find the marginal utility of leisure MU_F:

$$\frac{\partial U_A}{\partial F} = \frac{1}{3}(MF(18 - F)^{\frac{1}{3}-1}(18M - 2MF)$$

And then we differentiate with respect to M to find the marginal utility of consumption, MU_M:

$$\frac{\partial U_A}{\partial M} = \frac{1}{3}(MF(18 - F))^{\frac{1}{3}-1}(F(18 - F))$$

MRS then becomes:

$$MRS = \frac{MU_F}{MU_M} = \frac{M(18 - 2F)}{F(18 - F)}$$

At the optimum, the marginal rate of substitution (MRS) must equal the real wage w/p, which using $w = p = 1$ can be written as:

$$MRS = \frac{w}{p} \Rightarrow \frac{M(18 - 2F)}{F(18 - F)} = 1$$

This can be expressed as:

$$M = \frac{F(18 - F)}{(18 - 2F)}$$

We use the budget constraint $pM = w(18 - F)$, and the information given to us that $p = 1$, and insert in the optimality condition and get:

$$w(18 - F) = \frac{pF(18 - F)}{(18 - 2F)}$$

Which simplifies to:

$$F = 18 - 2F$$

This means that Audrey's choice of free time is given by:

$$F_A = 6$$

Her choice of time at the job, $J_A = T - F_A$, is therefore:

$$J_A = 18 - F_A = 12 \quad \text{Audrey's choice of work hours}$$

Audrey's labour supply is illustrated in Fig. 3.1b. Since Audrey enjoys working, she chooses to spend two-thirds of her available time working, 12 h per day, while, as you remember from the analysis above, Beth chooses to spend only half of her time working, 9 h.

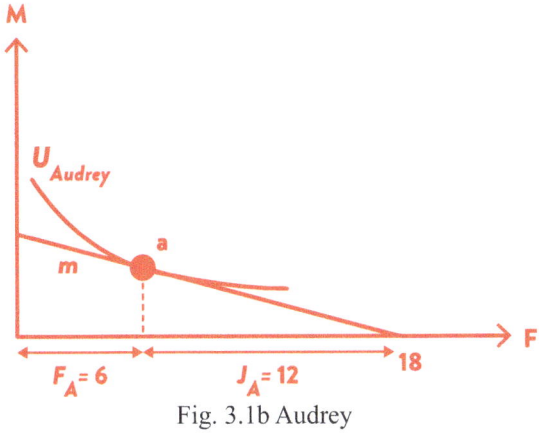

Fig. 3.1b Audrey

3.2 Reservation Wage

a. With a balanced Cobb-Douglas utility function, as in the exercise, that is $\alpha = 0.5$ and with $T = 16$, this means that as a freelancer you will divide your available time equally between leisure and work, $F_a = J_a = 8$.

With an hourly wage of $w_a = 2.25$, your income (and consumption) is $M_a = w_a J_a = 18$, and hence your utility is:

$$U_a = M_a^{0.5} F_a^{0.5} = (18)^{0.5}(8)^{0.5} = 12 \quad \text{Your utility from job a (freelance)}$$

b. As a manager, you must work $J_b = 12$ hours and thus have leisure $F_b = 4$, which gives you utility:

$$U_b = M_b^{0.5} F_b^{0.5} = (M_b)^{0.5}(4)^{0.5} \quad \text{Your utility from job b (manager)}$$

To be equally satisfied with the managerial job b as with your current job a, the following must hold:

$$U_b = U_a \Rightarrow (M_b)^{0.5}(4)^{0.5} = 12$$

Squaring both sides and rearranging, we get:

$$4M_b = (12)^2$$

$$M_b = 36$$

We see that you need an income increase of at least $M_b - M_a = 36 - 18 = 18$ to work the extra four hours ($J_b - J_a = 12 - 8 = 4$) required by the managerial position. Your reservation wage as a manager is therefore:

$$w_b = \frac{M_b}{J_b} = \frac{36}{12} = 3$$

This is illustrated in Fig. 3.2, where the higher reservation wage is shown by the steeper budget line m_2.

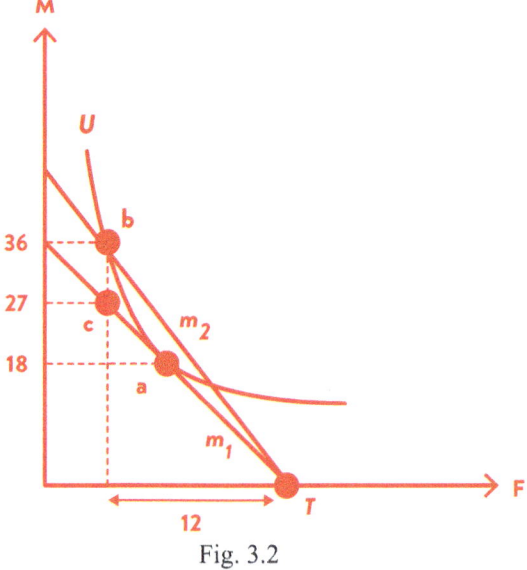

Fig. 3.2

The increase in income as a manager, from $M_a = 18$ to $M_b = 36$, is due to two factors: (i) more hours at work, and (ii) a higher hourly wage. The first effect can be found simply by starting with the original wage $w_a = 2.25$, which, with more hours worked, would give you an income $w_a J_b = 2.25(12) = 27$. This is marked as point c in Fig. 3.2. The difference $M_c - M_a = 27 - 18 = 9$ is due to working more hours.

The rest of the increase in income, that is, the difference $M_b - M_c = 36 - 27 = 9$ is therefore due to the higher hourly wage. In summary, half of your higher income comes from working more hours, and half from the higher hourly wage.

3.3 Labour Supply with Cash Transfer
a. With a cash transfer S, the budget constraint is:

$$w(T - F) + S = pM$$

And if we specify that $T = 16$ and $p = 1$, this can be written as:

$$w(16 - F) + S = M$$

We found in Math Box 3.1 that the optimal consumption combination can be expressed as:

$$M = \frac{F\alpha}{(1 - \alpha)} \frac{w}{p}$$

With a balanced Cobb-Douglas utility function ($\alpha = 0.5$) and with $p = 1$ this simplifies to:

$$M = wF$$

We substitute this expression into the budget constraint and get:

$$w(16 - F) + S = wF$$

which can be rearranged to:

$$2wF = 16w + S$$

and then:

$$F = 8 + \frac{S}{2w} \quad \text{Optimal choice of free time}$$

Since $J = T - F = 16 - F$, we can find the optimal labour supply as

$$J = 16 - \left(8 + \frac{S}{2w}\right) = 8 - \frac{S}{2w} \quad \text{Optimal choice of labour supply}$$

b. With $w = 1$ and $S = 0$, we have $J = 8$ (as expected, since a balanced Cobb-Douglas utility function results in an equal division of available time between work and leisure when all income comes from work), while for $S = 6$, we have $J = 5$. This is illustrated in Fig. 3.3.

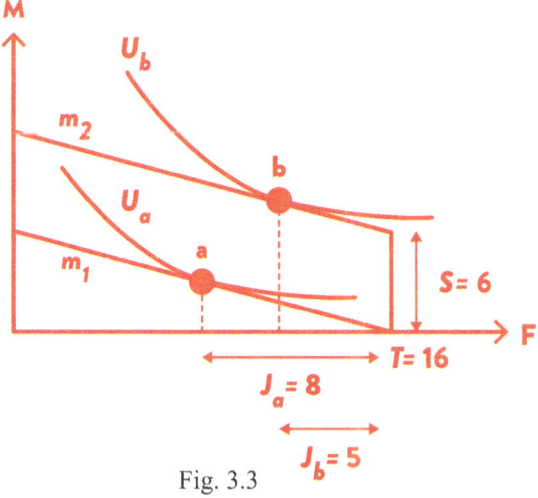

Fig. 3.3

Note that if the non-labour income becomes sufficiently large, the consumer will choose not to work at all—that is, a corner solution where $J = 0$. Substituting this into the expression for optimal labour supply, we get:

$$J = 0 \Rightarrow 8 = \frac{S}{2w}$$

From this, we can find the critical level of S that leads to the corner solution as:

$$S = 16w \quad \text{Critical level of S that gives zero labour supply}$$

This means that the higher the wage, the larger the cash gift from the grandfather must be for Anna to choose not to work.

3.4 The Shop Manager Thinks Anna Responds Strangely to a Wage Increase

a. In Exercise 3.3 above, we found that the labour supply with a cash transfer is given by:

$$J = 8 - \frac{S}{2w} \quad \text{Optimal choice of labour supply}$$

The first time the manager gave Anna a wage increase, she was not receiving any money from her grandfather ($S = 0$). From the expression for optimal labour supply, we see that in this case the labour supply is independent of the wage, $J = 8$ both before and after the wage increase. This happens because, with Cobb-Douglas preferences, the substitution and income effects cancel each other out.

Figure 3.4 illustrates this. The wage increase causes the budget line to rotate from m_1 to m_2, and the optimum moves from point a to point b, but this does not affect the labour supply, which remains at $J_a = 8$. The manager finds it strange that Anna does not want to work more for a higher wage, but this is because he does not see the income effect of the wage increase, which pulls towards lower labour supply.

b. The second time the manager gives her a higher wage, Anna is receiving support $S = 6$ from her grandfather, so the labour supply becomes $= 8 - 3/w$.

With the cash transfer, the wage increase causes the budget line in the figure to rotate from m_3 to m_4, and the optimum moves from point c to point d, increasing the labour supply from $J_c = 5$ to $J_d = 6$. Thus, with the cash transfer, the wage increase leads to higher labour supply, while without the cash transfer, the wage increase has no effect on labour supply.

This is what the manager originally expected—that the wage increase would lead to higher labour supply—but it contradicts what he saw the first time he increased the wage. He is confused.

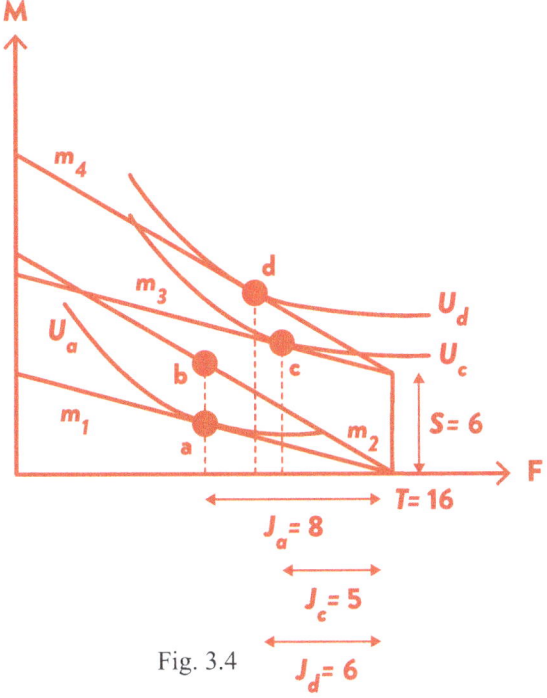

Fig. 3.4

c. It is easy to understand why the manager is confused, so how can we explain this?

Let's look again at the expression for the optimal labour supply:

$$J = 16 - \left(8 + \frac{S}{2w}\right) = 8 - \frac{S}{2w} \quad \text{Optimal choice of labour supply}$$

The higher the wage, the less important the second term become, that is, the less significance the non-labour income S has for the decision about labour supply (you can clearly see this by considering the extreme case when w approaches infinity, then the second term approaches zero). And the less important the non-labour income is, the more the consumer is drawn towards spending half of their time working, which is the "rule" for optimal labour supply with a balanced Cobb-Douglas utility function. With $T = 16$, this pulls the consumer towards $J = 8$.

Intuitively, this can be explained as follows. An increase in non-labour income always encourages more leisure. This follows because leisure and consumption are normal goods. Non-labour income produces a pure income effect. On the other hand, an increase in the wage has a more complicated effect on labour supply since it also contains a substitution effect, which encourages more work.

In general, we can say that the larger the share of your budget that comes from non-labour income, the less you want to work, while the larger the share that comes from wages, the more you want to work.

If we apply this insight to Fig. 3.4, starting from point c, a higher wage increases the relative importance of wages as a source of income compared to the non-labour income S, which makes you choose to work more, moving from $J_c = 5$ to $J_d = 6$.

3.5 Universal Basic Income and Labour Supply
a. We know from Math Box 3.1 that with the utility function $U = M^{0.5}F^{0.5}$ and $T = 16$, the labour supply is given by:

$$J = \alpha T = 8$$

With 8 h of work per day, consumption is:

$$M = wJ = 8$$

The allocation corresponds to point a in Fig. 3.5. Since leisure is $F = T - J = 8$, this allocation yields utility:

$$U_a = M^{0.5}F^{0.5} = 8^{0.5}8^{0.5} = 8$$

b. The benefit (welfare payment) makes it possible to consume $M_0 = 4$ even without working (which means $F = 16$), that is, point b in Fig. 3.5a. The utility in that case would be:

$$U_b = 4^{0.5}16^{0.5} = 8$$

Mark and Frank are therefore indifferent between working and living on welfare: $U_a = U_b \equiv U_{ab} = 8$. Working provides higher consumption but less leisure: without welfare, we have, $M = 8$ and leisure $F = 8$, while living on welfare gives consumption $M = S = 4$ and leisure $F = T = 16$.

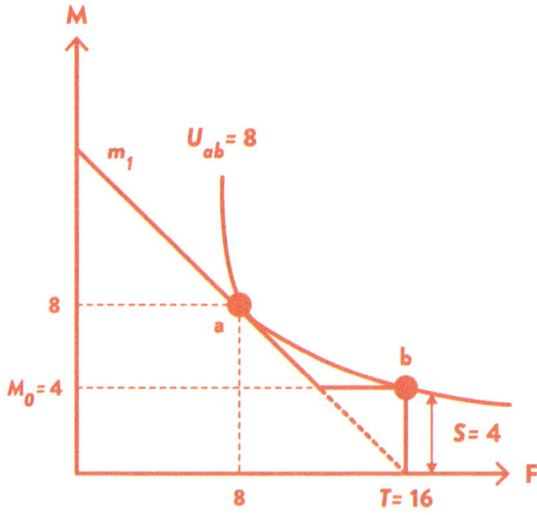

Fig. 3.5a

c. With an unconditional cash transfer (which we think of as a basic income) equal in size to the welfare Frank received, that is, $S = 4$, we know from Exercises 3.3 and 3.4 that the labour supply is given by:

$$J = \alpha T - \frac{(1 - \alpha)S}{w} = 0.5(16) - \frac{(1 - 0.5)4}{1} = 6$$

We see that the basic income causes Frank, who had previously been on welfare, to go from not working at all to working 6 h per day, a shift from point b to point c in Fig. 3.5b. A basic income may therefore be a sensible way to avoid a "welfare trap", where people like Frank choose welfare over work.

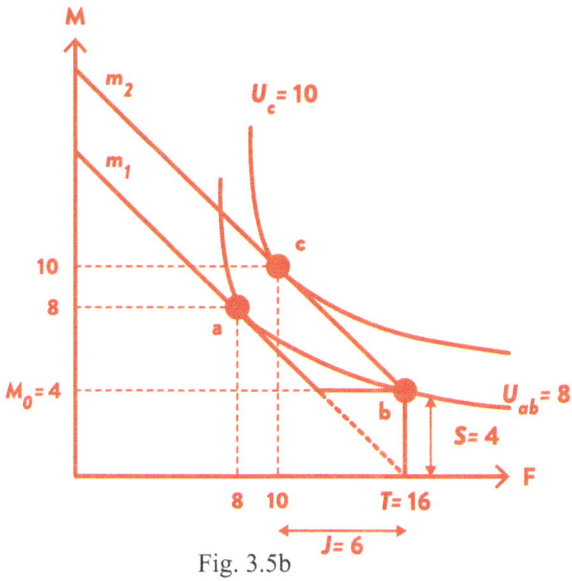

Fig. 3.5b

d. What about Mark? For him, the basic income means a shift from a to c: he chooses to work less when moving from welfare to basic income, from 8 h (as we found earlier in this exercise) to 6 h per day.

e. If we compare Frank and Mark, we see that the implications for labour supply from a transition from welfare to basic income are not clear-cut, as those who lived on welfare will enter the workforce, while those who were already working will work less.

Chapter 4 Behavioural Economics

4.1 Patient and Impatient Brian

a. Math Box 2.2 shows us how to find consumption over time with Cobb-Douglas preferences. With $I_1 = 1, I_2 = 0$, and, $r = 0$, consumption in period 1 is $x_1 = (1 - \alpha)I_1$, and consumption in period 2 is $x_2 = s = \alpha I_1$.

Patient Brian (Brian A) has $\alpha = 0.5$, and we see that he will choose to consume the same amount in both periods $x_1 = x_2 = s = 0.5$, which given the symmetry in preferences and the zero interest rate is quite intuitive. This is illustrated as point a in Fig. 4.1, the point where Brian A's indifference curve U_A^a is tangent to the budget line m.

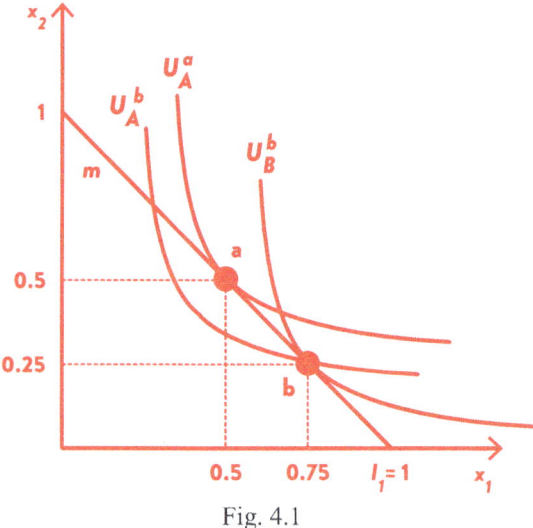

Fig. 4.1

b. Impatient Brian (Brian B) places greater weight on consumption today and has $\alpha = 0.25$. If there is no fee for withdrawing money from the savings account, he will choose consumption in the two periods as $x_1 = 0.75, x_2 = s = 0.25$.

This is marked as point b in the figure, the point where Brian B's indifference curve U_B^b is tangent to the budget line m. Brian B wants to spend more money on consumption today and would therefore, if allowed, withdraw some of the savings that Brian A has placed in the account.

c. Patient Brian will experience a utility loss if impatient Brian gets his way. At point a Brian A has a utility:

$$U_A^a = x_2^{0.5} x_1^{0.5} = (0.5)^{0.5}(0.5)^{0.5} = 0.5$$

While at point b, Brian A has utility:

$$U_A^b = x_2^{0.5} x_1^{0.5} = (0.25)^{0.5}(0.75)^{0.5} \approx 0.43$$

Accordingly, the utility loss for patient Brian of not being able to control his impatient self is:

$$U_A^a - U_A^b = 0.5 - 0.43 = 0.07$$

4.2 Brian Ties Himself to the Mast

a. Brian A has made an agreement with the bank that he must pay a fee g to withdraw money from the savings account—you can think of it as Brian imposing a penalty on himself! If Brian B uses the savings, then his income becomes $I_1 = 1 - g$. We know from the previous exercise that Brian A has deposited $s = 0.5$ in the savings account, and as long as he doesn't touch the savings, the budget line remains unchanged and is given by m in Fig. 4.2a. This applies to all points from a and upward (where a higher point on the budget line means saving even more). But to consume more today, $x_1 > 0.5$, he must dip into the savings. In that case, the budget line shifts inward by the fee g. In the figure this is shown by the budget line m'. The budget line thus has a kink, the original budget line applies for $x_1 < 0.5$, while the budget line reduced by the fee applies for $x_1 > 0.5$.

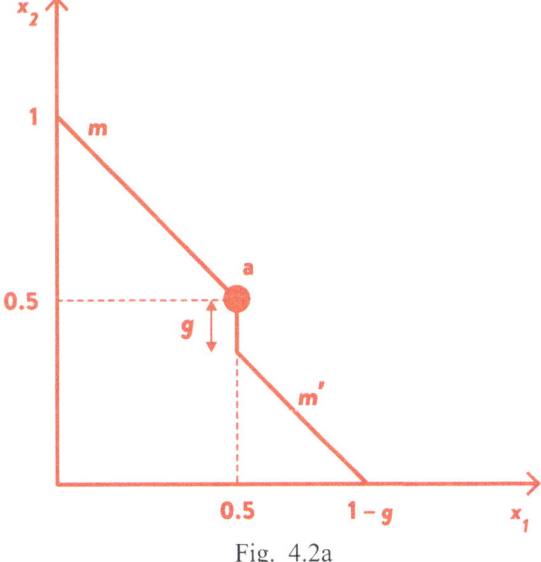

Fig. 4.2a

b. From the analysis above, we know that consumption in period 1 is given by $x_1 = (1 - \alpha)I_1$ and in period 2 by $x_2 = s = \alpha I_1$.

With a fee, income becomes $I_1 = 1 - g$, so if impatient Brian were to choose to withdraw money from the savings account, such that budget line m' would apply, then his optimal consumption would be:

$$x_1 = (1 - \alpha)I_1 = 0.75(1 - g)$$

$$x_2 = s = \alpha I_1 = 0.25(1 - g)$$

We now plug these values of x_1 and x_2 into Brian B's utility function and get:

$$U_B(g) = x_2^{0.25} x_1^{0.75} = \left[0.25(1 - g)\right]^{0.25} \left[0.75(1 - g)\right]^{0.75} \approx 0.57(1 - g)$$

c. Brian A's challenge is to impose a penalty on himself that is large enough to ensure that the impatient version of himself won't want to touch the savings. He therefore asks: How large must the fee be for Brian B to be indifferent between using the money and not using it (and thus, we assume, leave the savings untouched)? With this fee—or any higher—Brian A effectively ties himself to the mast.

The alternative for Brian B is not to touch the money in the savings account, which gives him equal consumption in both periods, $x_1 = x_2 = 0.5$. We insert this consumption into Brian B's utility function and find:

$$U_B^a = x_2^{0.25} x_1^{0.75} = (0.5)^{0.25} (0.5)^{0.75} = 0.5$$

This is impatient Brian's utility at point a in Fig. 4.2b. The fee that makes this version of Brian indifferent between withdrawing savings and leaving them in the account is found by:

$$U_B(g) = U_B^a \Rightarrow 0.57(1 - g) = 0.5 \Rightarrow g^* \approx 0.12$$

With such a fee, Brian, by using his savings, would end up at point c in Fig. 4.2b, with consumption $x_1 = 0.75(1 - g^*) \approx 0.66$ and $x_2 = 0.25(1 - g^*) \approx 0.22$. This consumption gives Brian B the same utility as at point a, where he does not touch the savings account. With a fee g^* or higher, Brian A has thus ensured that Brian B will leave the savings untouched!

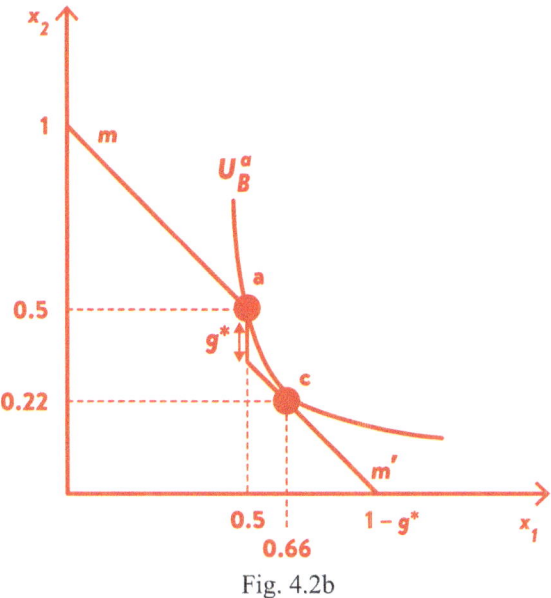

Fig. 4.2b

4.3 The Taxi Drivers in New York

a. We know that Allan, with a balanced Cobb-Douglas utility function, chooses labour supply:

$$J_A = T - F = \frac{1}{2}T$$

With $p = 1$, this means Allan's consumption is:

$$M_A = wJ_A$$

Inserting $T = 16$, we find that Allan will choose to split his time equally between work and leisure, $J_A = F_A = 8$, and thus his consumption will be $M_A = 8w$. Note that Allan's labour supply is independent of the wage: he always chooses to work 8 h a day.

Bill, on the other hand, has an income target, given as $M_B^* = 8$. He will therefore choose his labour supply to ensure this income target is met, meaning:

$$wJ_B = 8 \Rightarrow J_B = \frac{8}{w}$$

b. Scenario 1. Monday-Tuesday

We are told that the wage early in the week is low:

$$w_1 = \frac{2}{3} \quad \text{Low wage Monday and Tuesday}$$

This gives the budget line m_1, and Bill's labour supply is then:

$$J_B^1 = \frac{8}{w_1} = \frac{(3)8}{2} = 12 \quad \text{Bill's labour supply Monday and Tuesday}$$

His choice early in the week is shown as $Bill_1$ in Fig. 4.3a. Allan, on the other hand, always chooses an 8-h workday, as discussed in part (a), and therefore chooses $Allan_1$.

Scenario 2. Wednesday-Thursday
Midweek we have a somewhat higher wage:

$$w_2 = 1 \quad \text{Moderate wage Wednesday and Thursday}$$

This gives the budget line m_2, and Bill chooses a labour supply:

$$J_B^2 = \frac{8}{w_2} = \frac{8}{1} = 8 \quad \text{Bill's labour supply Wednesday and Thursday}$$

Midweek, the two drivers choose the same number of work hours, as indicated by $Bill_2$ and $Allan_2$ in Fig. 4.3b.

Scenario 3. Friday-Saturday
Late in the week we have:

$$w_3 = \frac{4}{3} \quad \text{High wage Friday and Saturday}$$

This gives the budget line m_3, and Bill works:

$$J_B^3 = \frac{8}{w_3} = \frac{(3)8}{4} = 6 \quad \text{Bill's labour supply Friday and Saturday}$$

This is shown as $Bill_3$ in Fig. 4.3c, while Allan chooses $Allan_3$.

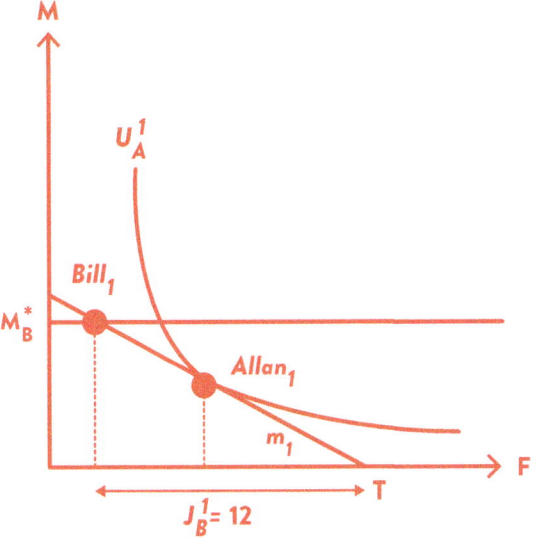

Fig. 4.3a Monday-Tuesday

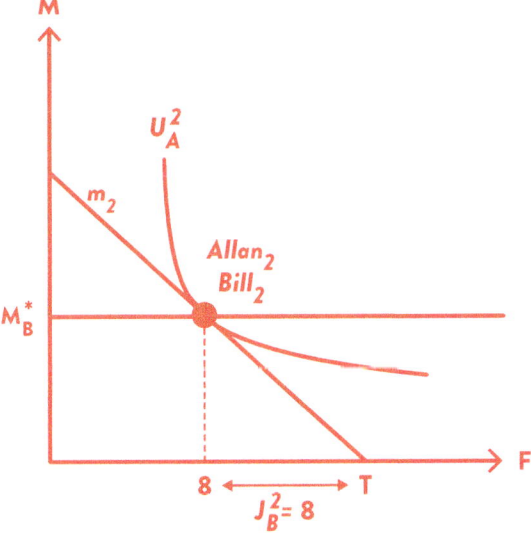

Fig. 4.3b Wednesday-Thursday

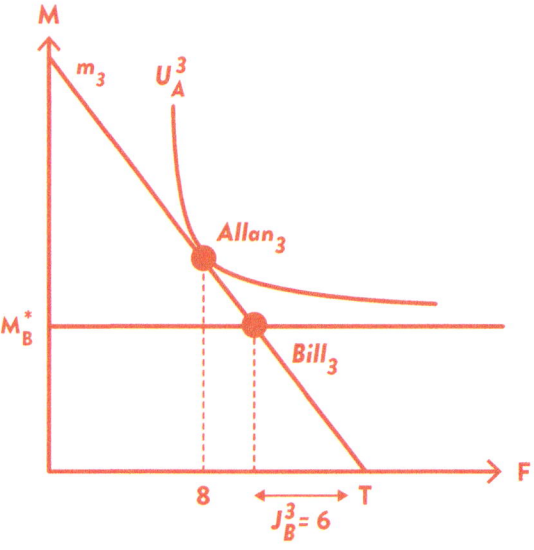

Fig. 4.3c Friday-Saturday

c. How much do they work, and what do they earn in a week? Allan works $J_A = 8$ every day, so the total number of hours worked in a week is:

$$\sum J_A = 6(8) = 48$$

Since his income is $M_A = wJ_A$, for example on Monday $M_A^1 = w_1 J_A$, his total income over the six working days is:

$$\sum M_{Allan} = 2\left(M_A^1 + M_A^2 + M_A^3\right) = 2(w_1 + w_2 + w_3)J_A$$
$$= 2\left(\frac{2}{3} + 1 + \frac{4}{3}\right)8 = 48$$

Bill works to meet his income target $M_B^* = 8$ each day, so his weekly income is simply:

$$\sum M_{Bill} = 6\left(M_B^*\right) = 6(8) = 48$$

That is, he earns the same as Allan. But how many hours does he work per week?

$$\sum J_B = 2J_B^1 + 2J_B^2 + 2J_B^3 = 2(12 + 8 + 6) = 52$$

So, Bill works more hours than Allan, but the total income is the same. This is because Bill works extra hours when the hourly wage is low (12 h on Monday and Tuesday).

4.4 Audrey's Altruism

a. From standard consumer theory, we know that the optimal choice between two goods is governed by the condition:

$$MRS = \frac{p_B}{p_A}$$

From Math Box 1.2, we know that with balanced Cobb–Douglas preferences ($\alpha = 0.5$) and equal prices for consumption goods, that is $p_A = p_B = p = 1$, this utility maximization condition can be written as:

$$\frac{A}{B} = 1$$

In other words, altruistic Audrey wants her son to have the same level of consumption as herself. Using the budget constraint $I_A = p_A A + p_B B$, which with price equal to 1 can be written as $I_A = A + B$ we find the optimal consumption for Audrey and Brian as:

$$A = B = 0.5$$

This is the optimal allocation initially, corresponding to point a in Fig. 4.6 in the textbook: $A_a = B_a = 0.5$. Here, the mother gives half of her income to her son, so they end up consuming the same amount. Quite intuitive, really, since she places equal weight on her son's consumption as on her own.

b. If Brian now earns his own income (I_B), he will pay for at least part of his own consumption. The mother only needs to pay for the son's consumption that exceeds what he earns himself, that is $B - I_B$, so Audrey's budget constraint can now be written as:

$$I_A = A + (B - I_B)$$

We use Audrey's optimal consumption combination found above, namely that $B = A$. This holds as long as the price remains unchanged, which it does:

$$I_A = A + (B - I_B) \Rightarrow I_A = A + (A - I_B) = 2A - I_B$$

We then use the fact that her income is $I_A = 1$ and Brian's income is $I_B = 0.25$, and substitute these values into the expression above:

$$I_A = 2A - I_B \Rightarrow 1 = 2A - 0.25$$

This gives:

$$A = 0.625$$

And since $A = B$:

$$B = 0.625$$

This corresponds to point c in Fig. 4.6 in the textbook: $A_c = B_c = 0.625$.

We see that while Audrey initially spent 0.5 on her son and 0.5 on herself, now that Brian earns money, she reduces the gift to $I_A - A_c = 1 - 0.625 = 0.375$, while Brian's consumption increases by $B_c - B_a = 0.625 - 0.5 = 0.125$, that is, his consumption increases by only half as much as his earnings. The reason lies in fungibility: when Brian earns his own money, Audrey reduces the gift to her son and uses the money on herself (as shown by the movement from point a to b in the figure in the textbook).

4.5 Give Smarter! Anna Offers Audrey Some Sound Advice

We start by drawing Brian's labour supply as the trade-off between consumption and leisure in the usual way, as shown in Fig. 4.5. Initially, before the gift, Brian has chosen point a on budget line m_1, with labour supply J_a and consumption M_a.

We know from Chapter 3 that an unconditional cash transfer lowers labour supply, as illustrated for example in Fig. 3.5 in the textbook. In Fig. 4.5 below, the cash transfer shifts the budget line outward to m_2, and Brian chooses point b, where he works J_b and consumes M_b.

The support from his mother finances most of his consumption, but not all of it. The job gives him an income at point d on budget line m_1, with income M_d, and the rest—that is, the difference between M_b and M_d—is the cash transfer from his mother: $M_b - M_d = S$.

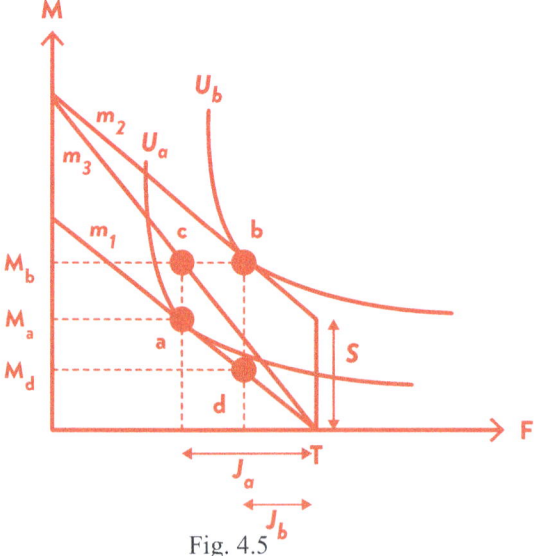

Fig. 4.5

Let us now consider Anna's suggestion of giving a wage subsidy. This makes the budget line steeper (as if Brian had received a higher wage). We know that with Cobb-Douglas preferences, labour supply is not affected by changes in the wage (see, for example, Fig. 3.4 in the textbook). The budget line m_3 gives consumption M_b with labour supply J_a. To keep the figure clear, I haven't drawn the indifference curve that is tangent to m_3, but since the labour supply is unchanged, his choice must be point c.

With a wage subsidy that gives the budget line m_3, Audrey thus achieves the same result as with the cash transfer: in both cases, consumption is M_b. But at a lower cost to the mother! With the wage subsidy, Brian's labour supply remains unchanged, so he finances a larger share of the consumption with his own earnings. Audrey's contribution is now $M_b - M_a$, which is considerably lower than the cash transfer S.

In sum, we see that while the lump sum transfer causes Brian to work less, the wage subsidy induces him to keep on working and in this way earn more of his own money.

Good advice, Anna!

Chapter 5 Labour and Capital

5.1 Same Quantity or Same Cost?
a. In Math Box 5.2, we see that the optimal factor combination for a balanced Cobb-Douglas production function is:

$$\frac{K}{L} = \frac{w}{r}$$

And factor demand is:

$$L = Q\sqrt{\frac{r}{w}}$$

$$K = Q\sqrt{\frac{w}{r}}$$

Moreoever, for $w = r = 1$ and output $Q = 1$, the optimal input mix becomes $L = K = 1$. This is illustrated in Fig. 5.1 as point a. The cost of production is then:

$$C_a = wL_a + rK_a = 2$$

This means the isocost line intersects the vertical axis where $L = 0 \Rightarrow C_a = rK = 2$, which with $r = 1$ gives $K = 2$, as shown in the figure. Similarly, it intersects the horizontal axis at $L = 2$.

b. The wage rate now increases to $w = 2$. What is the firm's optimal choice of inputs if it wants to keep output constant?

From the factor demand functions shown above, we see that for $Q = 1$ we get $L = Q\sqrt{\frac{r}{w}} = \sqrt{\frac{1}{2}} \approx 0.71$ and $K = Q\sqrt{\frac{w}{r}} = \sqrt{2} \approx 1.41$.

This is illustrated as point b in Fig. 5.1. The production cost is now (for simplicity using only one decimal in the factor inputs):

$$C_b = wL_b + rK_b = 2(0.7) + 1(1.4) = 2.8$$

c. What if the firm instead wants to keep production costs unchanged at the original level, i.e., $C = 2$?

We know from the optimal factor combination $K/L = w/r$ that with the higher wage rate $w=2$ (and $r=1$, as before), the firm should choose $K = 2L$. Using this and keeping costs constant at 2, we get $C_c = wL_c + rK_c = 2L + 1(2L) = 2 \Rightarrow L_c = 0.5$, which in turn implies $K_c = 1$. This is shown as point c in the figure.

The quantity produced in this case becomes $Q = K^{0.5}L^{0.5} = (1)^{0.5}(0.5)^{0.5} \approx 0.71$ as shown by the isoquant Q_2.

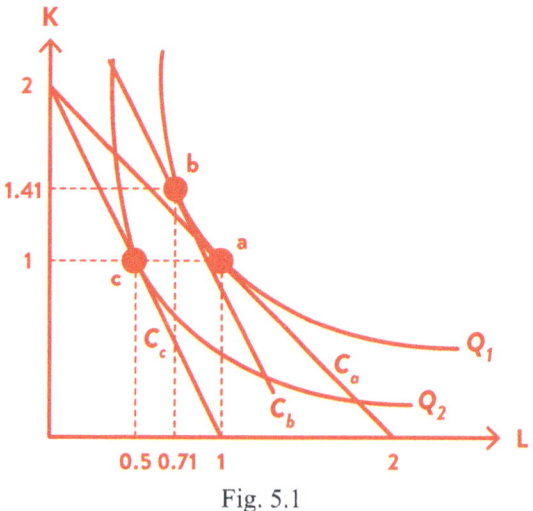

Fig. 5.1

The isocost C_c intersects the horizontal axis at $K = 0 \Rightarrow C_c = wL = 2$, which with $w = 2$, implies $L = 1$, as indicated in the figure.

5.2 How Much Does Conrad Save by Moving Production to China?

a. From Math Box 5.2, we know that the optimal factor combination for a balanced Cobb-Douglas production function is:

$$\frac{K}{L} = \frac{w}{r}$$

With factor prices at home, $w_H = r_H = 1$, we have $K = L$, and to produce $Q=1$, $K_H = L_H = 1$ as illustrated by the point a in Fig. 5.2. The total costs are therefore:

$$C_{Home} = w_H L_H + r K_H = 1(1) + 1(1) = 2$$

The wage in China is lower than at home, $w_C = 0.25$. If production is moved to China but uses the same factor combination as at home, i.e., $K_H = L_H = 1$, the cost becomes:

$$C'_{China} = w_C L_H + r K_H = 0.25(1) + 1(1) = 1.25$$

In Fig. 5.2, this means the production point is unchanged (same factor combination as at home), but at the new relative prices (lower wages in China mean a flatter isocost line), with costs C'_b. Without changing factor intensity in production, we see that Conrad saves $C_{Home} - C'_{China} = 2 - 1.25 = 0.75$ by moving production to China.

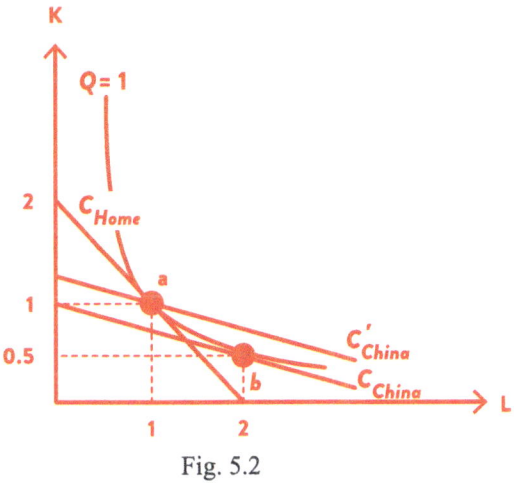

Fig. 5.2

b. Labour costs are reduced because of the move, but we see that Conrad can do better than this. The cost minimum is found where the isocost line is tangent to the isoquant, and for China this does not occur at point a.

With the factor price ratio in China, we know that:

$$\frac{K}{L} = \frac{w_C}{r} = 0.25 \Rightarrow K = 0.25L$$

Conrad should therefore use four times as much labour as capital in China to minimize costs. To produce $Q = 1$, this means:

$$Q = K^{0.5} L^{0.5} = 1 \Rightarrow (0.25L)^{0.5} L^{0.5} = 1 \Rightarrow L = 2$$

And since he should use four times as much labour as capital, this means $K = 0.5$. This is shown as point b in Fig. 5.2. With this factor usage, costs become:

$$C_{China} = w_C L + r_C K = 0.25(2) + 1(0.5) = 1$$

We see that production costs are halved by moving production to China: 75% of this saving comes without Conrad needing to think much—he can just move production to China and keep the same factor usage as at home while benefiting from lower wages. But 25% of the saving comes from Conrad thinking a bit: by optimizing factor usage according to the lower relative price of labour in China, he achieves an additional saving of $C'_{China} - C_{China} = 1.25 - 1 = 0.25$.

5.3 Leontief & Sons

a. The production functions in the three factories are (as the name implies) of the Leontief type, where we know that input factors must be combined in a fixed ratio, namely:

$$\frac{K}{\alpha} = \frac{L}{1 - \alpha}$$

With $Q = 1$ we therefore have:

$$\frac{K}{\alpha} = \frac{L}{1 - \alpha} = 1$$

$$K = \alpha \quad \text{Optimal choice of capital}$$

$$L = 1 - \alpha \quad \text{Optimal choice of labour}$$

This gives the cost of producing $Q = 1$:

$$C = wL + rK = w(1 - \alpha) + r\alpha$$

In the USA, factor prices are given by $w_U = r_U = 1$, so the costs are:

$$C_{USA} = w_U L + r_U K = 1(1 - \alpha) + 1(\alpha) = 1$$

That is, the same cost regardless of the technology parameter α. Figure 5.3a illustrates the choice of input combination in the USA on the isocost line C_{USA}, where Q_1 is the isoquant for the first factory ($\alpha_1 = 1/3$) and choice of point c; Q_2 is the isoquant for the second factory ($\alpha_2 = 0.5$) and choice of point b; and Q_3 is the isoquant for the third factory ($\alpha_3 = 2/3$) and choice of point a. We see that production in the third factory is the most capital-intensive, while the first factory is the most labour-intensive.

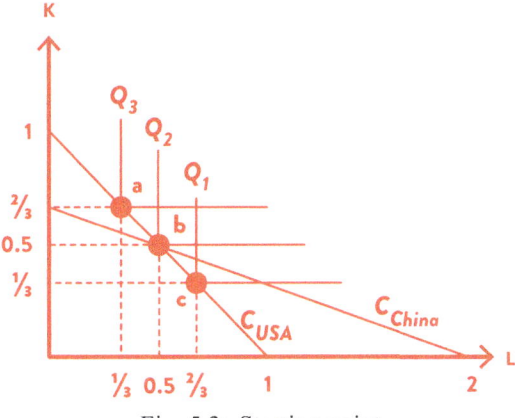

Fig. 5.3a Starting point

In China, factor prices are given by $w_C = 0.5, r_C = 1.5$, so production costs there become:

$$C_{China} = w_C L_C + r_C K_C = 0.5(1 - \alpha) + 1.5\alpha = 0.5 + \alpha$$

The cost of producing $Q = 1$ in the USA and China is equal when:

$$C_{USA} = C_{China} \Rightarrow 1 = 0.5 + \alpha \Rightarrow \alpha^* = 0.5$$

For $\alpha > \alpha^*$ then $C_{USA} < C_{China}$ while for $\alpha < \alpha^*$ we have $C_{USA} > C_{China}$. Factories with α higher than the critical level should therefore remain in the USA, while those with lower α should move to China.

Intuitively, a high α means high capital intensity in production, and since capital costs are higher in China, these firms should stay in the USA. For firms with low α, meaning more labour-intensive production, they should move to China where wages are lower.

Your advice would therefore be that the third Leontief factory should remain in the USA, while the first factory, the most labour-intensive one, should move to China, while for the second factory the costs are the same in both countries (so no gain from moving that one).

b. If the price of capital in China falls to $r'_C = 1.25$, your location advice must be updated. The cost of producing in China is now:

$$C'_{China} = w_C L + r'_C K = 0.5(1 - \alpha) + 1.25(\alpha) = 0.5 + 0.75\alpha$$

The cost of producing $Q = 1$ in the two countries is now the same when:

$$C_{USA} = C'_{China} \Rightarrow 1 = 0.5 + 0.75\alpha \Rightarrow \alpha' = \frac{2}{3}$$

Your moving advice to the Leontief family must be revised. We see that the cost for the third factory is now the same in both places, and this was a factory that you previously (with higher capital price) recommended to stay in the USA.

The new isocost line $C'_{China} = 1$ is illustrated in Fig. 5.3b. Now the two isocost lines intersect at point a, which is the chosen factor combination for the third factory, implying that for costs are the same in both locations for this factory, while the two other factories should move to China.

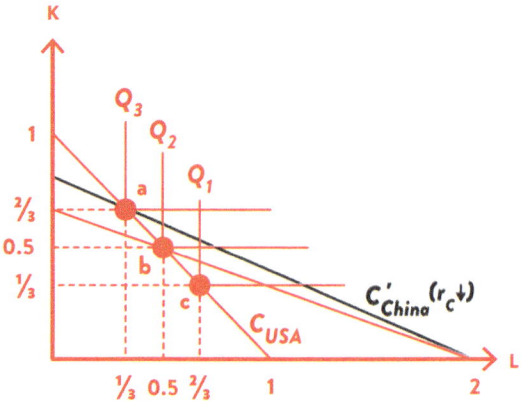

Fig. 5.3b Lower price of capital in China

5.4 On the Robotisation of Paper Production

a. With a balanced Cobb-Douglas production function $Q = K^{0.5}L^{0.5}$, we know that the optimal factor combination is:

$$\frac{K}{L} = \frac{w}{r}$$

With factor prices $w = r = 1$ and $Q = 1 \equiv Q_1$, the optimal factor input becomes $L = K = 1$, shown as point a in Fig. 5.4a. This gives total costs (using traditional technology) of $C = wL + rK = 2$, indicated by the isocost line C_a in the figure. This isocost line intersects the vertical axis where:

$$K = \frac{C_a}{r} \equiv K_1^a$$

We insert $C_a = 2$ and $r = 1$ and find that the cost expressed in capital units is:

$$K_1^a = 2$$

The capital requirement for robot production is given by:

$$R_1 = \frac{Q_1}{A}$$

With $A = 0.5$ and $Q_1 = 1$ this means:

$$R_1 = 2$$

So, it takes two units of robot capital to produce one unit of paper.

We see that the isocost in this case intersects the vertical axis at the same point as the capital requirement with robots, $K_1^a = R_1$. The costs are therefore the same, and there is no reason to switch technology.

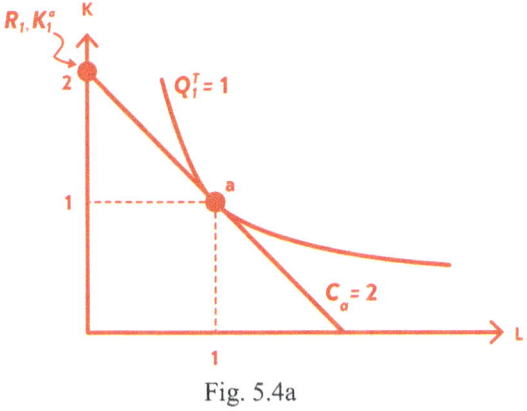

Fig. 5.4a

b. An increase in the price of capital to $r = 4$ causes the isocost to become flatter, as shown in Fig. 5.4b. The optimal factor combination is now:

$$\frac{K}{L} = \frac{w}{r} = \frac{1}{4} \Rightarrow K = 0.25L$$

Inserting this into the production function, and since one unit is to be produced, we get: $Q = K^{0.5}L^{0.5} = 1 \Rightarrow (0.25L)^{0.5}L^{0.5} = 1 \Rightarrow L = 2$, which implies that $K = 0.5$, which in Fig. 5.4b is illustrated as point b. The production costs are now:

$$C_b = wL + rK = (1)2 + (4)0.5 = 4$$

The new isocost intersects the vertical axis at:

$$K = \frac{C_b}{r} \equiv K_1^b$$

With $C_b = 4$ and $r = 4$ the costs expressed in units of capital now are:

$$K_1^b = 1$$

Since $R_1 = 2$ we now have $R_1 > K_1^b$ and robotisation would lead to higher costs than traditional production: don't do it!

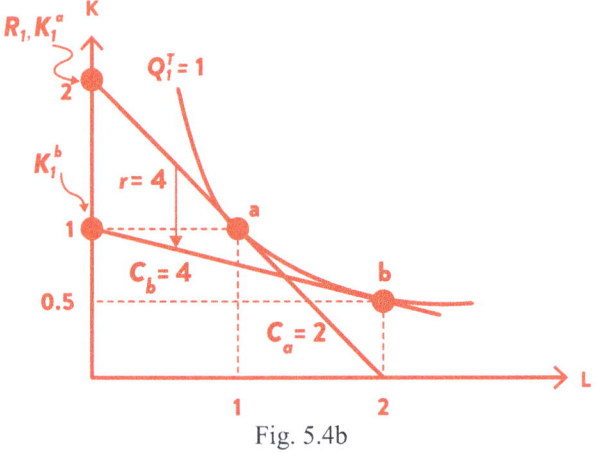

Fig. 5.4b

There are two reasons why the higher capital cost to a lesser degree affects traditional technology than it does robot technology. First, traditional technology uses *a mix* of capital and labour (and the price of labour is unchanged), so the increase in the price of capital only affects part of its cost base. Second, traditional technology allows for flexibility: firms can shift towards more labour-intensive production in response to higher capital costs, and this also reduces the impact of an increase in the price of capital on firms' costs. Robot technology does not allow for this kind of flexibility and is therefore more exposed to increases in the cost of capital.

c. We see from the analysis above that $K_1^b = 1$ with traditional technology and the high price of capital. What does it take to make robotisation profitable? To answer this question, we start by finding the quality level of the robots that makes the firm indifferent between robotising and keeping the traditional technology. The costs of the two technologies are the same when $R_1^{new} = K_1^b$, which again means that:

$$R_1^{new} = K_1^b \Rightarrow \frac{Q_1}{A^{new}} = 1 \Rightarrow \frac{1}{A^{new}} = 1 \Rightarrow A^{new} = 1$$

For $A^{new} = 1$, we get $R_1^{new} = 1$, and the firm will be indifferent between the two technologies. A robot technology with $A^{new} > 1$ is therefore required for the firm to choose robotisation when the capital cost is as high as $r = 4$.

5.5 Robots or China?
a. Home
With robot technology, production is given by:

$$Q = AK = 0.75K$$

For $Q_1 = 1$ the required amount of robot capital is:

$$R_1 = \frac{Q_1}{A} = \frac{4}{3}$$

What about traditional technology? From Math Box 5.2, we know that the optimal factor combination for a balanced Cobb-Douglas production function is

$$\frac{K}{L} = \frac{w}{r}$$

With factor prices at home $w_H = r_H = 1$, we get $K = L$, and one unit of each factor is used to produce $Q_1 = 1$, as illustrated at the point a in Fig. 5.5. Factor usage is therefore $K_H = L_H = 1$. The total cost is:

$$C_{Home} = w_H L_H + r_H K_H = 1(1) + 1(1) = 2$$

The isocost line C_{Home} crosses the vertical axis at:

$$K = \frac{C_H}{r_H} \equiv K_1^a$$

With $C_{Home} = 2$ and $r_H = 1$ we get:

$$K_1^a = 2$$

The cost of traditional production at home, measured in units of capital, is 2.

Since we know that $R_1 = \frac{4}{3}$, this means that $R_1 < K_1^a$. Thus, Conrad should choose to automate production if he stays at home.

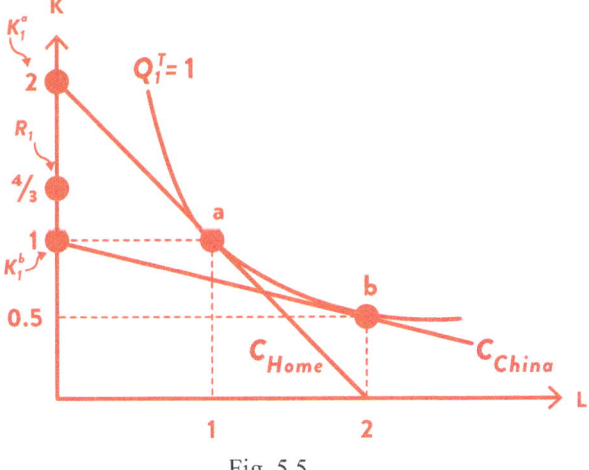

Fig. 5.5

b. China

With factor prices $w_C = 0.5, r_C = 2$, we know that the optimal factor combination using traditional technology in China is given by:

$$\frac{K}{L} = \frac{0.5}{2} = \frac{1}{4}$$

So, in China, the firm will choose to use four times as much labour as capital with traditional technology $K = 0.25L$. Substitute this optimal factor ratio into the production function $Q = 1$:

$$Q = (0.25L)^{0.5}L^{0.5} = 1 \Rightarrow L_C = 2$$

Since $K = 0.25L$, we then get $K_C = 0.5$. This corresponds to the point b in Fig. 5.5.

The total costs using traditional technology in China are:

$$C_{China} = w_C L_C + r_C K_C = 0.5(2) + 2(0.5) = 2$$

With the isocost line C_{China}, the cost expressed in capital units is:

$$K = \frac{C_C}{r_C} \equiv K_1^b$$

Vi substitute for $C_{China} = 2$ and $r_C = 2$ and get:

$$K_1^b = 1$$

We see that $R_1 > K_1^b$, so Conrad should choose the traditional technology if he moves production to China.

c. Home or China?

If Conrad chooses to stay at home, he should robotise, requiring robot capital $R_1 = \frac{4}{3}$, which with $r_H = 1$ gives us costs:

$$C_{Home} = r_H R_1 = (1)\frac{4}{3} = \frac{4}{3}$$

If he chooses China, he should use traditional technology, and as shown above, the cost is:

$$C_{China} = 2$$

Conrad should therefore stay at home and automate production, and not move to China.

Chapter 6 Costs

6.1 The Seamstress in Dar es Salaam

a. This task explores the concept of opportunity cost. Grace owns two sewing machines and her house, including the kitchen she uses as production space. She doesn't pay herself a wage but earns money from selling what she makes.

From Grace's perspective, all production inputs, including her own labour, are costless, since she's not paying for them. But as economists, we must consider opportunity costs: the value of the production inputs in their best alternative use.

We are told that there is a market for renting out sewing machines, which means that both the one she uses and the one that sits unused in the cupboard have an alternative value—what she could earn by renting them out. Similarly, Grace could likely earn a wage working as a seamstress for someone else, which implies that her time also has an opportunity cost—the wage she foregoes by being self-employed.

What about the production space, the kitchen? The key question is whether it could be rented out for other production purposes. The exerciset provides no information about this, but it seems unlikely that others would consider Grace's kitchen for production (or that she would embrace that idea, for that matter). In that case, the opportunity cost of the kitchen is zero. Economically, this makes the kitchen a good workspace for her.

Taken together, Grace must account for the cost of her own labour and the value of both sewing machines, measured by their market value (how much she could earn by working for others or by renting out the sewing machines). If she ignores this, she underestimates the cost of her production, and her home-based kanga business is less profitable than she believes.

b. Moving production out of the house and into rented facilities leads to higher costs, even in economic terms (that is, measured by opportunity cost). This is because, as mentioned above, there is no realistic alternative use for the kitchen that would generate income, which effectively makes it free capital that it makes sense to use as a production site. Shifting production to another space would therefore increase costs.

c. A good piece of advice for Grace is to rent out the extra sewing machine she has at home. Another sensible recommendation is to continue using her home as a workplace, if she plans to keep producing clothes herself. Ultimately, whether she should continue producing clothes depends on whether the economic profit—revenues minus all opportunity costs—from running her own business exceeds what she could earn as an employee elsewhere.

6.2 Visiting Alpha Tech

a. Math Box 6.1 gives us the marginal cost and average total cost for a balanced CD-production function, and using the information that $w = r = 1, z = 0$, we get:

$$MC = \frac{2wQ}{K_0} + z = \frac{2Q}{K_0}$$

$$ATC = \frac{wQ}{K_0} + z + \frac{rK_0}{Q} = \frac{Q}{K_0} + \frac{K_0}{Q}$$

During your first visit, $K_0 = 1$, and the two cost functions can be written as:

$$MC_1 = 2Q \quad \text{Marginal cost, first visit}$$

$$ATC_1 = \frac{wQ}{K_0} + z + \frac{rK_0}{Q} = Q + \frac{1}{Q} \quad \text{Average total cost, first visit}$$

These cost functions are illustrated in Fig. 6.2.

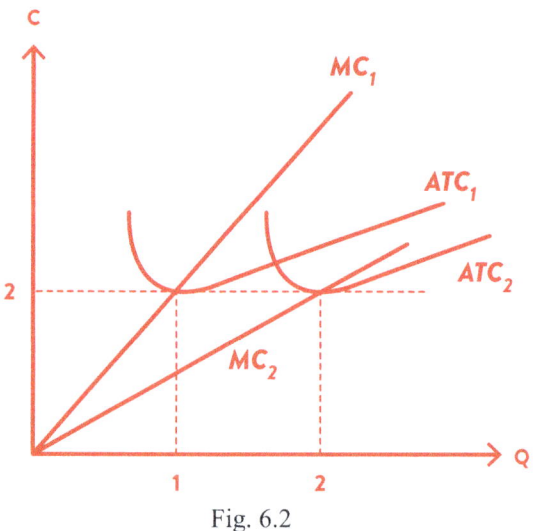

Fig. 6.2

b. During your second visit, $K_0 = 2$, and the two cost functions then become:

$$MC_2 = Q \quad \text{Marginal cost, second visit}$$

$$ATC_2 = \frac{Q}{2} + \frac{2}{Q} \quad \text{Average total cost, second visit}$$

These cost functions can also be found in Fig. 6.2. Relative to the first visit, the marginal cost increases at only half the rate.

Compared to your first visit, we note that the average total cost decreases over a larger range of production. This is due to a combination of two factors. First, a higher fixed cost makes it beneficial to increase production to spread those costs over more units. Second, more capital increases labour productivity, reducing the importance of labour as a cost driver.

c. You argue that the key reason behind the lower marginal cost during your second visit is tied to the labour input requirement. From Math Box 6.1 we know that:

$$L = \frac{Q^2}{K_0}$$

During your first visit $K_0 = 1$, and the labour requirement is $L_1 = Q^2$, while during your second visit, with $K_0 = 2$, it is $L_2 = 0.5Q^2$, that is only half of what you found during your first visit. By making labour more productive, the higher level of capital has significantly reduced the labour requirement for increased production.

6.3 Leontief & Sons Considers Expansion
a. We are told that each of the factories have installed the optimal level of capital to produce one unit of shirts ($Q = 1$), which from Exercise 5.3 we know means $K = \alpha$.

Variable costs are those associated with the variable factor, that is, labour: $VC = wL$. To express this in terms of units of output, Q, we need to use the production function:

$$Q = min\left(\frac{K}{\alpha}, \frac{L}{1 - \alpha}\right)$$

As long as the installed capital is not a binding constraint, production is determined only by the input of labour, that is, $Q = L/(1 - \alpha)$. We can reorganize this and write it in terms of labour requirement as $L = (1 - \alpha)Q$. Plugging this into the expression for variable costs, and using the fact that $w = 1$, we find that:

$$VC = wL = w(1 - \alpha)Q = (1 - \alpha)Q$$

We now have variable costs as a function of quantity Q. Marginal costs up to the capacity constraint (that is, up to $Q = 1$), are therefore:

$$MC = \frac{\partial VC}{\partial Q} = 1 - \alpha$$

The marginal costs of the three brothers are therefore:

$$MC_1 = 2/3 \quad \text{First factory}\left(\alpha_1 = \frac{1}{3}\right)$$

$$MC_2 = 1/2 \quad \text{Second factory} \left(\alpha_2 = \frac{1}{2} \right)$$

$$MC_3 = 1/3 \quad \text{Third factory} \left(\alpha_3 = \frac{2}{3} \right)$$

The marginal costs of the brothers' shirt production are illustrated in the figure below.

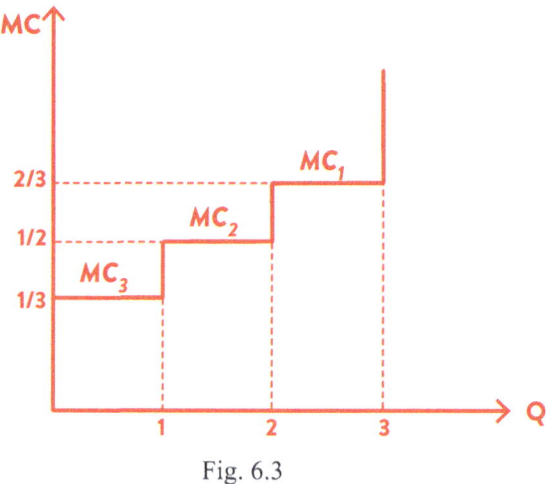

Fig. 6.3

b. Given the production technology at Leontief & Sons, with its strict complementarity between capital and labour, it is in fact not possible to expand production beyond three units, as indicated by the vertical marginal cost curve at $Q = 3$ in the figure. At this production level, the total installed capital in the three factories becomes a binding constraint, and further expansion would require additional capital investment.

6.4 A Green Transition in the Cardboard Industry

a. The total cost with brown production is:

$$C_B = wL + zQ + rK_0$$

We use the given production function to find the labour requirement as a function of output (as shown in Math Box 6.1 and during your visit to Alpha Tech):

$$L = \frac{Q^2}{K_0}$$

Since $K_0 = 1$ and the wage is $w = 1$, the labour cost can be written as $wL = Q^2$. Electricity cost is $Z = zQ$, which means that the variable costs with brown production are:

$$VC_B = Q^2 + zQ$$

With $r = 1$ and $K_0 = 1$, the fixed costs become $FC_0 = rK_0 = 1$, and the total costs—consisting of variable costs (labour and electricity) and fixed costs (capital)—can be written as:

$$C_B = VC_B + FC_0 = Q^2 + zQ + 1$$

A green transition involves an additional fixed cost $FC_G = 3$ for the incineration facility, but it eliminates the need for electricity for heating. The total costs then become:

$$C_G = wL + FC_0 + FC_G = Q^2 + 4$$

The marginal costs in the two cases are found by differentiating the cost function with respect to output Q (as shown in Math Box 6.1):

$$MC_B = \frac{\partial C_B}{\partial Q} = 2Q + z$$

$$MC_G = \frac{\partial C_G}{\partial Q} = 2Q$$

We find the average total cost by dividing total cost by output Q:

$$ATC_B = \frac{C_B}{Q} = Q + z + \frac{1}{Q}$$

Likewise, after a green transition, the firm has:

$$ATC_G = \frac{C_G}{Q} = Q + \frac{4}{Q}$$

b. We can find the production quantity that gives the lowest average total cost by taking the derivative of ATC_B and setting it equal to zero:

$$\frac{\partial ATC_B}{\partial Q} = 1 - \frac{1}{Q^2} = 0 \Rightarrow Q = 1$$

Alternatively, you could find the minimum point by setting $MC = ATC$ —you can verify for yourself that this gives the same result.

Note that the factory using the brown technology reaches the lowest cost at $Q = 1$, regardless of the level of z.

We then insert $Q = 1$ into ATC_B and find that:

$$ATC_B = \frac{C_B}{Q} = Q + z + \frac{1}{Q} = 1 + z + \frac{1}{1} = 2 + z$$

Similarly, we find the minimum point of the average total cost after the green transition as:

$$\frac{\partial ATC_G}{\partial Q} = 1 - \frac{4}{Q^2} = 0 \Rightarrow Q = 2$$

We insert this value into ATC_G and get:

$$ATC_G = \frac{C_G}{Q} = Q + \frac{4}{Q} = 2 + \frac{4}{2} = 4$$

Earlier, we found that $ATC_B = 2 + z$. We therefore have:

$$ATC_B = ATC_G \Rightarrow 2 + z = 4 \Rightarrow z = 2$$

This is illustrated in the figure below, which shows the case where $z = 2$ and thus the same minimum ATC for the two technologies.

If the electricity price exceeds this threshold, the cost curve for brown technology shifts upward, making it the more costly technology. Conversely, a lower electricity price will shift the brown costs downward, making this the cheaper technology.

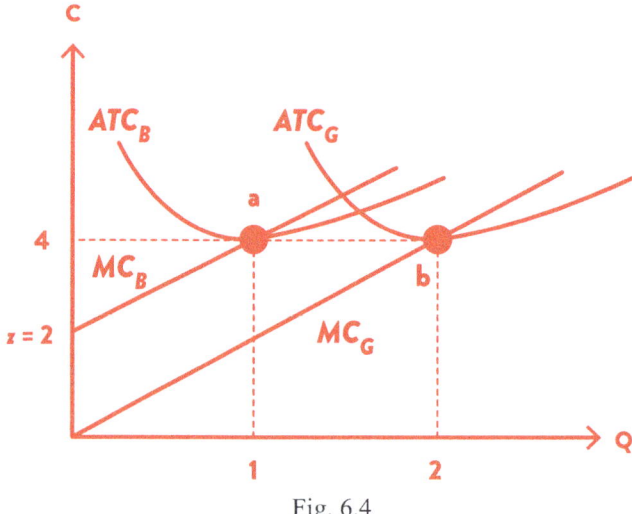

Fig. 6.4

6.5 Conrad's Two Factories

In the South, costs are given by $C_S = Q_S^2$. Since we're looking at a short-term time horison, we should think of these as variable costs. The marginal costs can be found as:

$$MC_S = \frac{\partial C_S}{\partial Q_S} = 2Q_S$$

And average variable costs:

$$AVC_S = \frac{C_S}{Q_S} = Q_S$$

At the factory in the North, costs are given by $C_N = 0.6Q_N$ which means that the marginal costs there are:

$$MC_N = \frac{\partial C_N}{\partial Q_N} = 0.6$$

Which in this case is identical to the average cost:

$$AVC_N = \frac{C_N}{Q_N} = 0.6$$

a. Brian's proposal is to equalise average costs, $AVC_N = AVC_S$. Since $AVC_S = Q_S$ while $AVC_N = 0.6$, this implies $AVC_S = AVC_N \Rightarrow Q_S = 0.6$, and since $Q = 1$ we therefore have $Q_N = Q - Q_S = 1 - 0.6 = 0.4$. In the figure below, Brian's preferred point is b. The total costs from the two factories in this case would be:

$$C_{Brian} = AVC_S Q_S + AVC_N Q_N = 0.6(0.6) + 0.6(0.4) = 0.6$$

The costs under Brian's proposal are illustrated by area S for the costs in the South and N for the costs in the North.

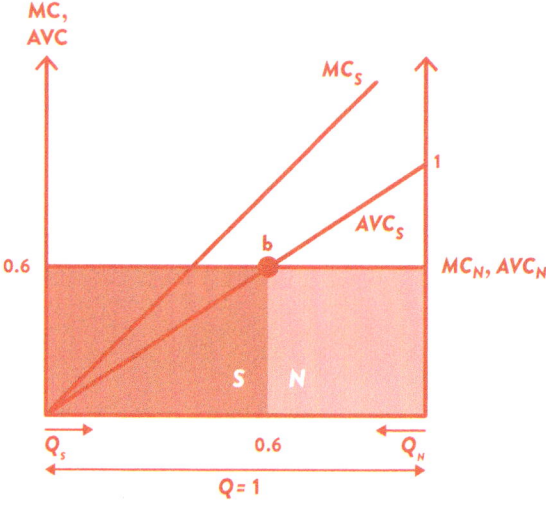

Fig. 6.5a Brian's proposal

b. Conrad proposes to split production equally between the factories in the North and South, that is, $Q_N = Q_S = 0.5$, marked as point c in the figure below. This yields the following costs:

$$MC_N = AVC_N = 0.6$$

$$MC_S = 2Q_s = 1$$

$$AVC_S = Q_s = 0.5$$

The total costs from the two production sites when production is organised in this way are given by:

$$C_{Conrad} = AVC_S Q_S + AVC_N Q_N = 0.5(0.5) + 0.6(0.5) = 0.55$$

The costs under Conrad's proposal are illustrated by the areas S and N in Fig. 6.5b.

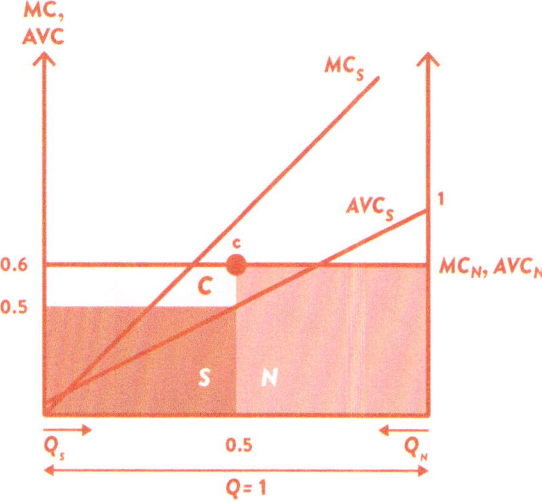

Fig. 6.5b Conrad's proposal

We note that Conrad's proposal results in lower costs than Brian's. The difference, which we calculated as $C_{Brian} - C_{Conrad} = 0.05$ corresponds to the area C in Fig. 6.5b.

c. Anna emphasizes the importance of marginal analysis. By this, she means that the marginal cost should be equal in both factories. Neither Brian's nor Conrad's proposals achieve this: in both cases, $MC_S > MC_N$ so equalising the marginal costs implies reducing production in the South (where marginal cost increases with output) and increasing production in the North (where marginal cost is constant).

Since $MC_S = 2Q_S$ and $MC_N = 0.6$, setting these equal gives $MC_N = MC_S \Rightarrow 2Q_S = 0.6 \Rightarrow Q_S = 0.3$, which means the remaining production is allocated to the Northern plant $Q_N = Q - Q_S = 1 - 0.3 = 0.7$, marked as point a in Fig. 6.5c.

Under Anna's proposal, costs are:

$$C_{Anna} = AVC_S Q_S + AVC_N Q_N = 0.3(0.3) + 0.6(0.7) - 0.51$$

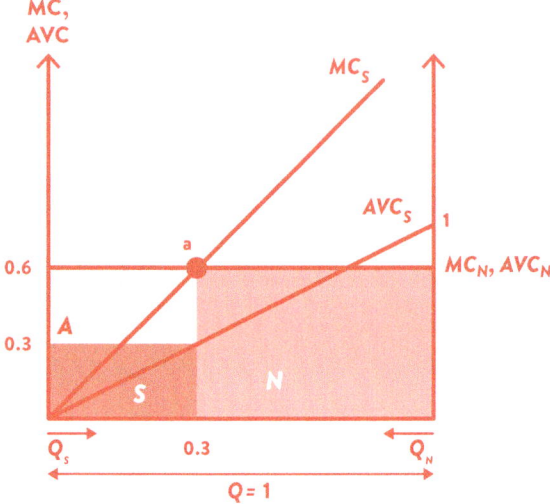

Fig. 6.5c Anna's proposal

We see that Anna's proposal yields the lowest total costs. The cost saving compared to Brian's proposal is given by area A, with size $C_{Brian} - C_{Anna} = 0.09$.

Anna's proposal is also better than Conrad's, as area A is larger than C, with the difference:

$$C_{Conrad} - C_{Anna} = 0.04.$$

Note that the loss from deviating from the optimal rule $MC_S = MC_N$ can alternatively be illustrated as the area between the marginal cost curves of the two factories, as shown in Fig. 6.5 in the textbook.

Triangle C in that figure corresponds to the cost increase from moving from Anna's solution to Conrad's $C_{Conrad} - C_{Anna} = 0.04$, while the triangle formed by the sum of areas B and C in Fig. 6.5 gives the additional cost of choosing Brian's solution $C_{Brian} - C_{Anna} = 0.09$.

Chapter 7 Profit

7.1 Profits in the Green Cardboard Factory

a. In Exercise 6.4, we studied the costs associated with a green transition in the cardboard industry. Now we turn to profitability. We know that the level of production that maximises profits is given by $P = MC$. With costs $C_G = Q^2 + 4$, we know that marginal costs are:

$$MC = \frac{\partial C_G}{\partial Q} = 2Q$$

With the given price $P_1 = 4$ and the marginal cost function we found there, we can easily determine the optimal quantity as

$$P_1 = MC_G \Rightarrow 4 = 2Q \Rightarrow Q_a = 2$$

This corresponds to point a in Fig. 7.1a. We see that at this point, the price equals the average total cost, and profits are therefore zero. This means the firm is breaking even, so there is no reason to shut down production, but also no reason to invest in new capacity.

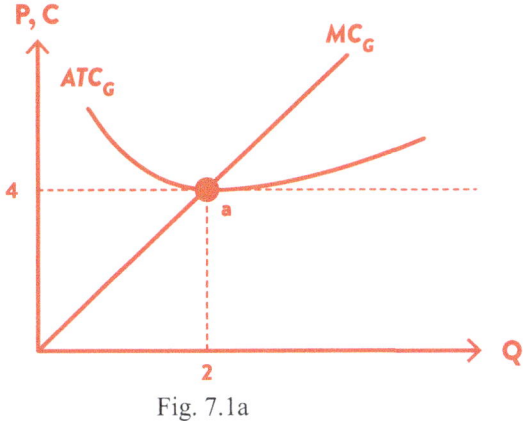

Fig. 7.1a

b. With costs $C_G = Q^2 + 4$, we know that if production remains unchanged at $Q_a = 2$, then $C_G = (2)^2 + 4 = 8$. With the price increasing to $P_2 = 6$, but with no change in production profits are:

$$\pi_b = P_2 Q_a - C_G = (6)(2) - 8 = 4$$

This is shown as the coloured rectangle in Fig. 7.1b.

c. However, keeping production constant is not the profit-maximizing choice at the higher price. The firm should choose a level of production such that marginal cost equals price:

$$P_2 = MC_G \Rightarrow 6 = 2Q \Rightarrow Q_c = 3$$

Production should therefore increase from 2 to 3 (point c), as shown in Fig. 7.1c. Profits are again shown as a shaded rectangle.

With $Q_c = 3$, costs become $C_G = (3)^2 + 4 = 13$. In that sense, the critics were right: higher production leads to higher costs. But revenue also increases, and with it, profits:

$$\pi_c = P_2 Q_c - C_G = (6)3 - 13 = 5$$

As you can see, $\pi_c - \pi_b = 5 - 4 = 1$.

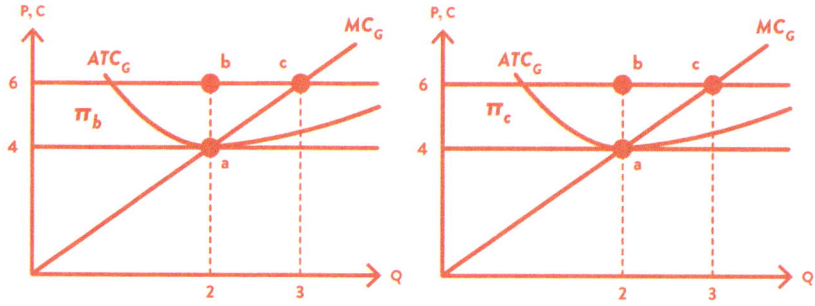

Fig. 7.1b Production unchanged.   Fig. 7.1c Optimal production

7.2 Profits Go Up and Down

From the discussion of Fig. 7.1 in the textbook, we know that point a represents a situation where, at the price P_2, profits are zero because average total costs are very high (driven by fixed costs); then profits increase as output rises, and we reach point b, which is the profit-maximising quantity; beyond that, profits start to fall again, and we reach point c where profits are once again zero—this time driven by high variable costs; and finally point d, where the operating profits are zero. In this exercise, we use a specific production function and a numerical example to find the production levels corresponding to these points in the figure.

Let us start with the profit maximum, point b. The marginal cost using the numbers given in the exercise ($w = r = z = K_0 = 1$) is:

$$MC = 2Q + 1$$

The condition for profit maximisation is that price equals marginal cost, which gives:

$$P = MC \Rightarrow P = 2Q + 1$$

Solving for quantity, we get (see also Math Box 7.1):

$$Q = 0.5(P - 1)$$

With $P = 5$, this means that the optimal quantity is:

$$Q = 0.5(5 - 1) = 2$$

This is point b in Fig. 7.1 in the textbook, and a similar graph is shown below as Fig. 7.2, including the relevant production quantities from the exercise.

At point a, profits are zero at a low output level. The formula for this is found in Math Box 7.1, which, together with $P = 5$, gives:

$$Q = \frac{P - 1 - \sqrt{(P-1)^2 - 4}}{2} = \frac{5 - 1 - \sqrt{(5-1)^2 - 4}}{2} \approx 0.27$$

This corresponds to the quantity produced at point a. Similarly, we find point c as the higher quantity where profits are zero.

$$Q = \frac{P - 1 + \sqrt{(P-1)^2 - 4}}{2} = \frac{5 - 1 + \sqrt{(5-1)^2 - 4}}{2} \approx 3.73$$

This corresponds to the quantity produced at point c. Finally, we find point d where the operating profits are zero, and according to Math Box 7.1, this is given by $Q = P - 1$, which with $P = 5$ gives $Q = 5 - 1 = 4$.

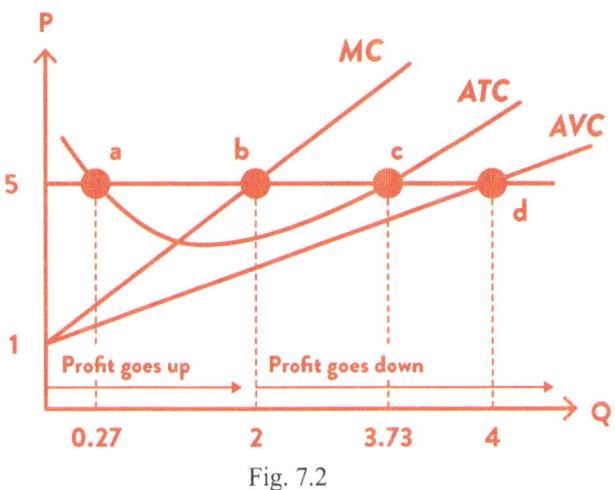

Fig. 7.2

In summary: If we start with $Q = 0$ and increase production, profits first rise and then fall. With $P = 5$, profits are negative for $Q < 0.27$; zero at $Q = 0.27$; reaching a maximum at $Q = 2$; then declining and zero again at $Q = 3.73$, becoming negative beyond this point.

7.3 More or Less Satisfied Capital Owners
a. Initially, the price is $P = 5$, and as we found in the previous exercise, the producer chooses $Q = 2$. With this quantity, we know from Math Box 7.1 that profits are:

$$\pi = PQ - C = PQ - (Q^2 + Q + 1) = 5(2) - (2^2 + 2 + 1) = 3$$

In Fig. 7.3, the profits are shown as the coloured area $\pi = (P - ATC)Q$.

b. We start by finding the quantity that gives the lowest average total cost, ATC, defined as:

$$ATC = \frac{C}{Q}$$

With $C = Q^2 + Q + 1$, this means that:

$$ATC = \frac{Q^2 + Q + 1}{Q} = Q + 1 + \frac{1}{Q}$$

These costs reach their minimum point when the following condition is met:

$$\frac{\partial ATC}{\partial Q} = 1 - \frac{1}{Q^2} = 0 \Rightarrow Q = 1$$

We also know that the marginal cost (MC) curve intersects the ATC curve at this minimum point, that is, at $Q = 1$, as illustrated at point b. Since the marginal cost in the example is given by $MC = 2Q + 1$, the marginal cost at $Q = 1$ is $MC = 2(1) + 1 = 3$.

Therefore, $P = 3$ is the critical price for the firm. At this price, the firm will choose $Q = 1$ (since marginal cost then equals price), and this equilibrium gives a profit $\pi = 0$. A price lower than this, that is $P < 3$, would lead to $\pi < 0$, and over time the firm should shut down production (since the capital could yield higher returns elsewhere).

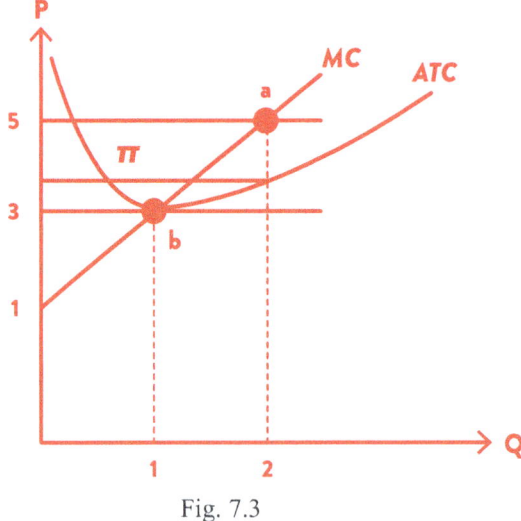

Fig. 7.3

7.4 Higher Gas Prices: Should We Shut Down the Factory?

Figure 7.4 illustrates a situation where the gas price is initially z_1, the price of the good produced is P, and the factory chooses an output level at point a, thus earning a profit (since $P > ATC$ at the optimal output).

An increase in the price of natural gas shifts both the marginal cost (MC) curve and the average total cost (ATC) curve upward (see Math Box 6.1). We know that the MC curve intersects the ATC curve at its minimum point, and at gas price z_2, the optimal output moves to point b, where $P = ATC$.

At this gas price, the factory has zero profits. Any gas price higher than this will result in negative profits, which in the long run will make it optimal for the firm to shut down production.

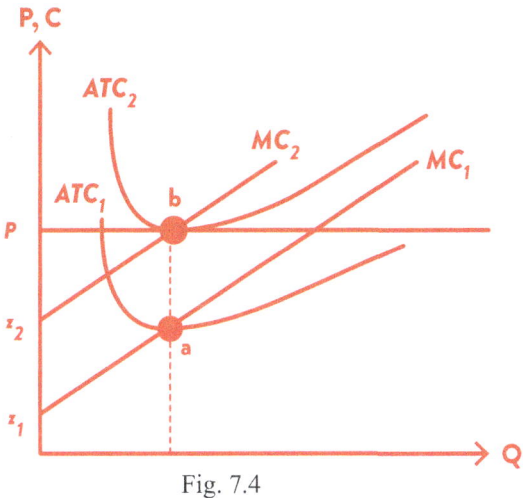

Fig. 7.4

7.5 The Firm's Demand for Labour

a. For the production function $Q = K^{0.5}L^{0.5}$, we know from Math Box 5.1 (by setting $\alpha = 0.5$) that the marginal product of labour (MP_L) is:

$$MP_L = 0.5 \left(\frac{K}{L}\right)^{0.5}$$

We use the first-order condition for profit maximisation, $PMP_L = w$:

$$P\left(0.5\left(\frac{K_0}{L}\right)^{0.5}\right) = w$$

Using the fact that the installed capital is $K_0 = 1$, this can be written as:

$$\frac{2w}{P} = \left(\frac{1}{L}\right)^{0.5} \Rightarrow 4\left(\frac{w}{P}\right)^2 = \frac{1}{L}$$

Rearranging to find labour demand L:

$$L = \frac{1}{4}\left(\frac{P}{w}\right)^2 \quad \text{Demand for labour}$$

With $P_{low} = 2$ labour demand becomes:

$$L = \frac{1}{4}\left(\frac{P_{low}}{w}\right)^2 = \frac{1}{4}\left(\frac{2}{1}\right)^2 = \frac{1}{w^2}$$

This is illustrated as labour demand curve L_1^D in Fig. 7.5. For a wage $w_{high} = 1$, labour demand is $L = 1$ (shown at point a). For a lower wage, $w_{low} = 0.5$, labour demand is $L = 4$, as shown at point b.
With $P_{high} = 4$ labour demand becomes:

$$L = \frac{1}{4}\left(\frac{P_{high}}{w}\right)^2 = \frac{1}{4}\left(\frac{4}{1}\right)^2 = \frac{4}{w^2}$$

This is illustrated as labour demand curve L_2^D in the figure. For $w_{high} = 1$, labour demand is $L = 4$ (shown at point c).

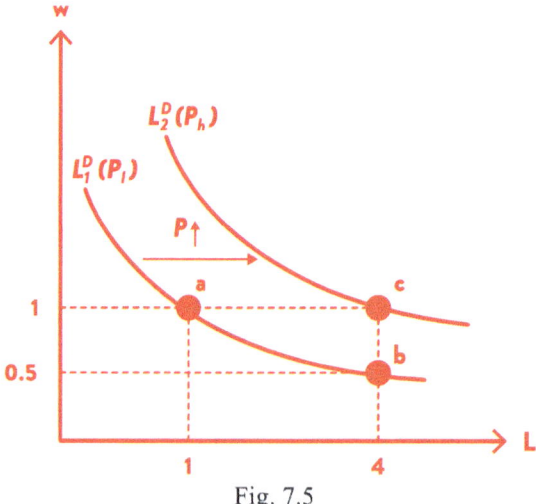

Fig. 7.5

Chapter 8 Perfect Competition

8.1 Demand for Gas from Households and Industry
a. The choke price for consumers is given by $P_D^{choke} = \alpha/\beta$.

For households, the demand function is $Q_H^D = 5 - P$. This means that $\alpha_H = 5$ and $\beta_H = 1$, which implies $P_H^{choke} = 5$.

For industry, the demand function is $Q_I^D = 6 - 2P$. This means that $\alpha_I = 6$ and $\beta_I = 2$, so that $P_I^{choke} = 3$. This is illustrated in Fig. 8.1a.

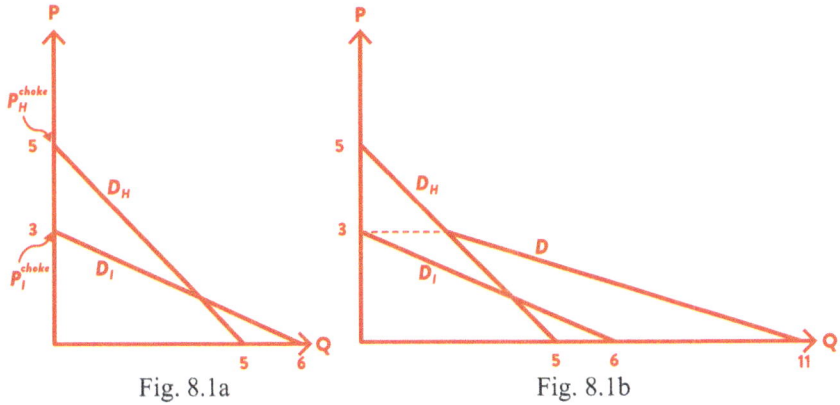

Fig. 8.1a   Fig. 8.1b

b. The total market demand is found by summing the two demand functions. Note that the aggregate demand curve has a kink at the industry's choke price, $P_I^{choke} = 3$. For prices above this level, only households demand gas.

$$Q^D = Q_H^D + Q_I^D = 5 - P + 6 - 2P = 11 - 3P \text{ for } P \leq 3$$

$$Q^D = Q_H^D = 5 - P \text{ for } P > 3$$

This is illustrated in Fig. 8.1b.

c. $P_I = 1$ gives total demand $Q^D = 8$, as shown in Fig. 8.1d, evenly divided between households, $Q_H = 4$, and industry, $Q_I = 4$, as shown in Fig. 8.1c.

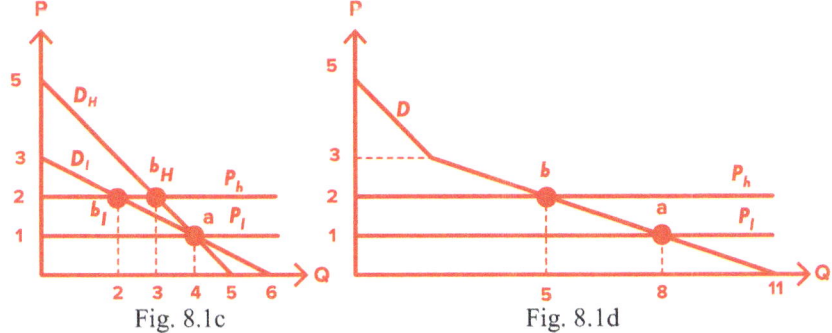

Fig. 8.1c   Fig. 8.1d

If the gas price rises to $P_h = 2$, we move from point a to point b in Fig. 8.1d, where total demand falls from 8 to 5.

We see that the increase in the gas price leads to the largest reduction in demand from industry, from 4 to 2, as shown in Fig. 8.1c by the shift from point a to point b_I. Households, on the other hand, reduce their demand from 4 to 3, moving from point a to point b_H.

The stronger response to the price increase from industry can be explained by the fact that the demand from the industry is more price sensitive than that of households.

8.2 War and Gas Prices in the Short and Long Run

a. The demand curves in the short and long run are illustrated in the figure as D_{short} and D_{long}. We know that the choke price for demand is given by:

$$P_D^{choke} = \frac{\alpha}{\beta}$$

In the short run, where $\alpha = \beta = 1$, the choke price is:

$$P_{short}^{choke} = 1$$

In the long run, where $\alpha = 1.5, \beta = 2$, we get:

$$P_{long}^{choke} = 0.75$$

This is illustrated in Fig. 8.2 as the points where the demand curves intersect the vertical axis. The demand curves intersect the horizontal axis at $Q = \alpha$, which means $\alpha_{short} = 1$ in the short run and $\alpha_{long} = 1.5$ in the long run.

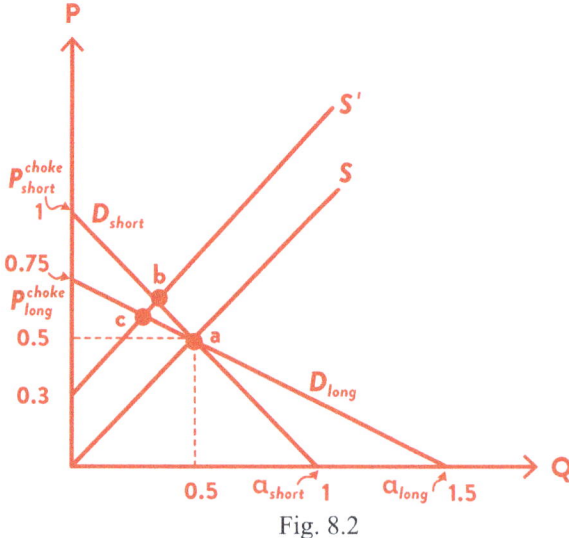

Fig. 8.2

b. From Math Box 8.3, we know that the equilibrium price is given by:

$$P^* = \frac{\alpha + \gamma}{\beta + \delta} \quad \text{Equilibrium price}$$

In this exercise, the gas supply is given by $Q^S = -\gamma + P$, which means $\delta = 1$.

In the short run, we have $\alpha_{short} = \beta_{short} = 1$, so before the invasion of Ukraine, where $\gamma = 0$, the equilibrium price becomes:

$$P_{short} = \frac{\alpha_{short} + \gamma}{\beta_{short} + \delta} = \frac{1 + 0}{1 + 1} = 0.5$$

And in the long run:

$$P_{long} = \frac{\alpha_{long} + \gamma}{\beta_{long} + \delta} = \frac{1.5 + 0}{2 + 1} = 0.5$$

As we can see, the price is the same in the short and long run in the absence of a supply shock.

From Math Box 8.1, we know that the price elasticity of demand is given by:

$$\varepsilon^D = \frac{\alpha - Q^D}{Q^D}$$

Both in the short and long run, $Q^D = 0.5$ in equilibrium. Since $\alpha_{short} = 1$, we get the short-run price elasticity of demand:

$$\varepsilon^D_{short} = \frac{\alpha_{short} - Q^D}{Q^D} = \frac{1 - 0.5}{0.5} = 1$$

In the long run, with $\alpha_{long} = 1.5$, we get:

$$\varepsilon^D_{long} = \frac{\alpha_{long} - Q^D}{Q^D} = \frac{1.5 - 0.5}{0.5} = 2$$

We see that demand is more elastic in the long run than in the short run.

c. With the supply shock given by $\gamma' = 0.3$, we get a new short-run equilibrium at point b in Fig. 8.2, with the new equilibrium price:

$$P'_{short} = \frac{\alpha_{short} + \gamma'}{\beta_{short} + \delta} = \frac{1 + 0.3}{1 + 1} = 0.65$$

The long-run equilibrium is at point c with price:

$$P'_{long} = \frac{\alpha_{long} + \gamma'}{\beta_{long} + \delta} = \frac{1.5 + 0.3}{2 + 1} = 0.6$$

We see that the supply shock leads to a smaller price increase in the long run than in the short run. This is because demand is more elastic in the long run than in the short run: a more elastic demand means that the cost increase resulting from the loss of Russian gas has less impact on the price, since consumers are better able to shift away from gas to alternative energy sources and to invest in energy-saving measures.

8.3 Pennies from Heaven

a. In the dry year, $V_l = 0.75$, the limiting factor is the availability of water, not the capacity of the power plants (which is given as $K = 1$). The supply is therefore given by S_1 in Fig. 8.3, with a vertical supply curve at the capacity limit $V_l = 0.75$, and zero marginal cost before that.

Summer demand, D_l, intersects the horizontal axis ($P = 0 \Rightarrow Q_l^D = 0.5$) to the left of V_l. Even though it has rained little this year, there is more than enough water in the reservoirs to meet summer demand, and the electricity price is zero, since the water is free. This is illustrated by point a in Fig. 8.3.

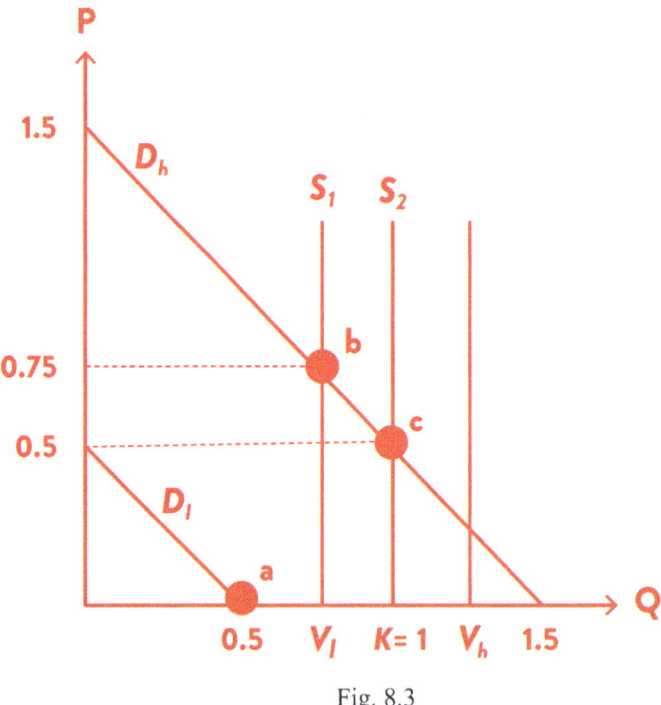

Fig. 8.3

In winter, demand D_h intersects the horizontal axis ($P = 0 \Rightarrow Q_h^D = 1.5$) to the right of the capacity limit $V_l = 0.75$. This would imply a water shortage, which drives the price up. The equilibrium price is found where demand equals supply:

$$Q_h^D = Q^S \Rightarrow 1.5 - P = 0.75 \Rightarrow P = 0.75$$

This is illustrated by point b in Fig. 8.3. The price is much higher than production costs, because the price balances supply and demand. So, while the water is free, it has rained little, and electricity becomes expensive due to high winter demand.

b. In the wet year, $V_h = 1.25$, and now the limiting factor is the production capacity, $K = 1$. The supply is therefore given by S_2 in the figure, with a vertical supply curve at the capacity limit $K = 1$, and zero marginal cost before that.

Summer demand, D_l, intersects the horizontal axis ($P = 0 \Rightarrow Q_l^D = 0.5$) to the left of $K = 1$. So again, the electricity price is zero in summer, as shown by point a in the figure.

In winter, demand D_h, intersects the horizontal axis ($P = 0 \Rightarrow Q_h^D = 1.5$) to the right of the capacity limit $K = 1$. This would again imply a shortage—this time defined by the size of the generation capacity— which drives the price up. The equilibrium price is found where demand equals supply:

$$Q_h^D = Q^S \Rightarrow 1.5 - P = 1 \Rightarrow P = 0.5$$

This is illustrated by point c in the figure. Again, the price is much higher than the production cost, but not as high as during the winter in the dry year.

8.4 Climate Change and International Trade in Agriculture
a. Market equilibrium occurs where supply intersects demand. In a normal year, we find the market equilibrium as

$$Q_{norm}^S = Q^D \Rightarrow -20 + P = 100 - P \Rightarrow P_{norm} = 60$$

which implies $Q_{norm} = 40$. We mark this as point a in Fig. 8.4a.

In a bad year, the market equilibrium is

$$Q_{bad}^S = Q^D \Rightarrow -40 + P = 100 - P \Rightarrow P_{bad} = 70$$

which implies $Q_{bad} = 30$. We mark this as point b in the figure. Thus, the negative weather shock has resulted in higher prices and lower production and consumption of agricultural goods at home.

In a good year, the market equilibrium is

$$Q_{good}^S = Q^D \Rightarrow P = 100 - P \Rightarrow P_{good} = 50$$

which implies $Q_{good} = 50$. We mark this as point c in the figure. We see that the positive weather shock has led to lower prices and higher production and consumption of agricultural goods.

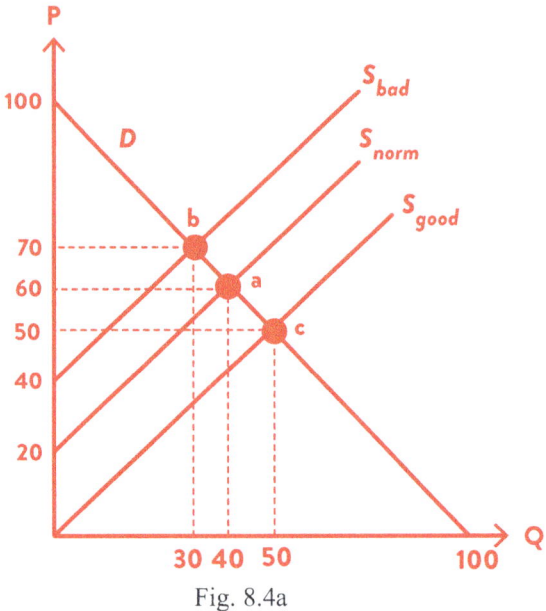

Fig. 8.4a

b. We now move on to the scenario with international trade at a world price $P_{int} = 60$. In a normal year, this will not affect the market equilibrium relative to the autarky solution described in part (a), since the autarky price in the normal year, $P_{norm} = 60$, is identical to the international price. In Fig. 8.4b the equilibrium is still at point a.

What about a bad year? Since the international price is not affected by changes in the weather in the home country, neither is consumption, which remains at point a with $Q^D = 40$. But the same is not true for the local producers. Insert $P_{int} = 60$ into the supply function $Q^S_{bad} = -40 + P_{int} \Rightarrow Q^S_{bad} = 20$, marked with point d in Fig. 8.4b. The difference beween consumption and local production is covered by imports: $Q^{imp}_{bad} = Q^D - Q^S_{bad} = 40 - 20 = 20$ Finally, let's consider a good year. Now insert $P_{int} = 60$ into the supply function $Q^S_{good} = P_{int} \Rightarrow Q^S_{good} = 60$, marked with point e in Fig. 8.4b. Production is now larger than consumption and the country becomes an exporter: $Q^{exp}_{good} = Q^S_{good} - Q^D = 60 - 40 = 20$.

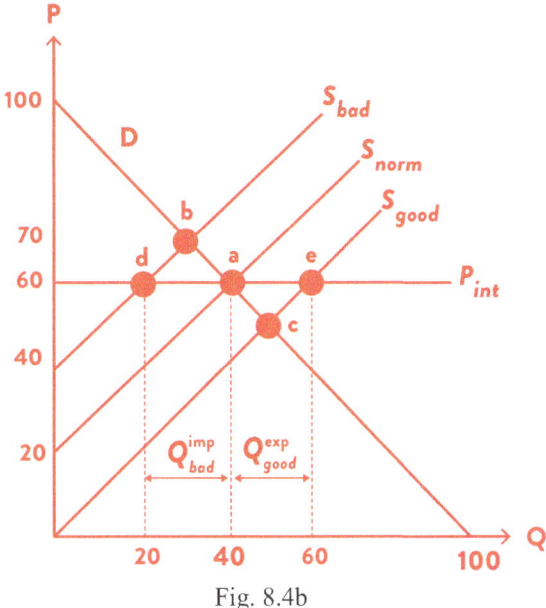

Fig. 8.4b

c. Let's first consider a bad year. Without trade, droughts and floods lead to a higher price and both production and consumption decline. With international trade, however, the price is unaffected by changes in local weather conditions, which is good news for consumers: they can continue to consume at the same price as before the weather shock. For producers, however, a bad year now implies a large loss of production, as imports replace a significant share of local production.

Now consider a good year for local agriculture. Without trade, this would lead to a lower price and increased consumption and production. With international trade, however, consumers are unaffected by the positive weather shock: they still have to pay the international price. In contrast, producers gain a lot, as they can sell their produce on the international market at the high, international price.

Accordingly, who wins from international trade, consumers or producers, depends on the weather!

8.5 Anna Points Out That Trade Can Create New Growth Opportunities for the Mill
Scenario 1
This scenario shows a negative demand shock, as illustrated in Fig. 8.5a. Initially, we have equilibrium at point a in autarky, where the demand curve D_1 intersects the supply curve S. We assume that this market equilibrium gives a price under autarky that is equal to the international price, P_{int}, so production in both cases is Q_a.

Now we assume a fall in domestic demand, illustrated by the shift from D_1 to D_2. In autarky, this means a new equilibrium at point b, with lower production, Q_b.

But what happens when there is trade? In that case, production stays at Q_a. Demand now lies at point c, and the difference between production and domestic demand $(Q_a - Q_c)$ is exports.

We note that trade in this scenario makes Conrad and the other paper producers less vulnerable to a decline in domestic demand: They can always choose to export instead!

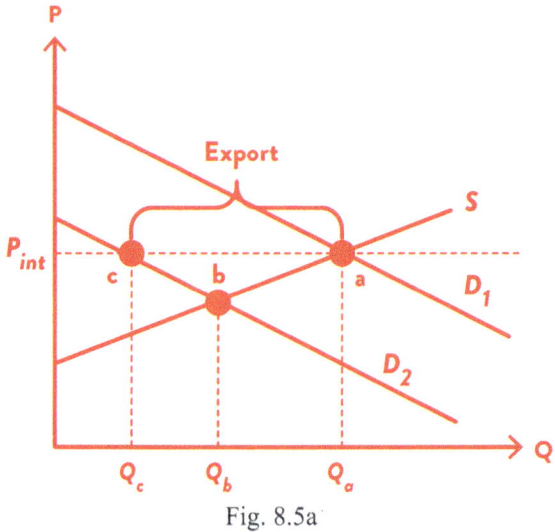

Fig. 8.5a

Scenario 2

This scenario considers a positive supply shock, for instance due to lower energy prices, and is illustrated in Fig. 8.5b. The starting point is a, with production Q_a. The lower energy price leads to a positive shift in the supply curve, from S_1 to S_2. Under autarky, this means a new equilibrium at point b, with higher production, Q_b.

But with international trade, the effect of lower production costs for production is even greater. We move from point a to point c, with production Q_c. Demand remains at point a, and the difference between production and domestic demand $(Q_c - Q_a)$ is exports.

Note that under autarky, the shift in the supply curve leads to a lower price, which dampens the profitability of expanding production. With trade, however, producers can sell as much as they want at the international price, and therefore the lower electricity price results in a larger increase in production in this case.

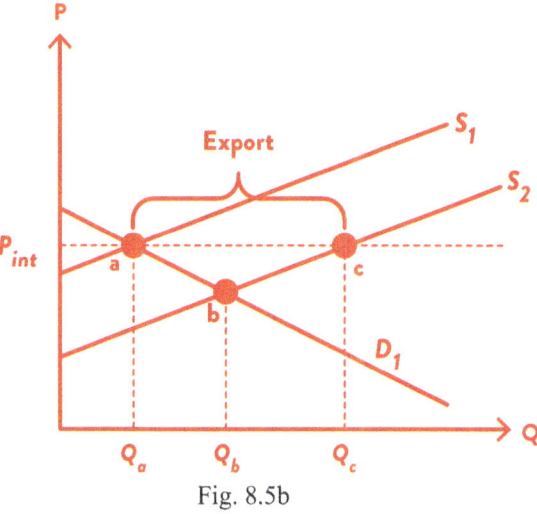

Fig. 8.5b

Scenario 3

We here investigate the impact of a higher international price, illustrated in Fig. 8.5c. The starting point is again a, with production Q_a. But with international trade, the rise in the international price, to P'_{int}, leads to a movement from point a to point c for producers, and from point a to point b for consumers, and where the difference between production and domestic demand ($Q_c - Q_b$) is exports.

Once again, we see an example of how international trade can lead to an increase in production.

Of course, opening up for trade does not always increase production. What happens, for example, if the international price falls below the autarky price?

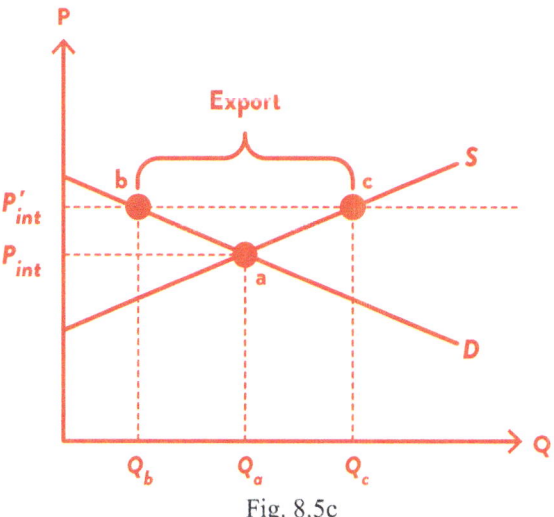

Fig. 8.5c

Chapter 9 Economic Efficiency

9.1 War on the Continent, Higher Gas Prices and Economic Surplus

a. We find the market equilibrium (see Math Box 8.3) where $Q^D = Q^S$, which with the given functions means $1 - P = -\gamma + P$, leads to the equilibrium price:

$$P^* = \frac{1 + \gamma}{2}$$

For $\gamma_{peace} = 0$, we have $P_{peace} = 0.5$, which means $Q_{peace} = 0.5$, i.e., the equilibrium is at the point marked "peace" in Fig. 9.1a.

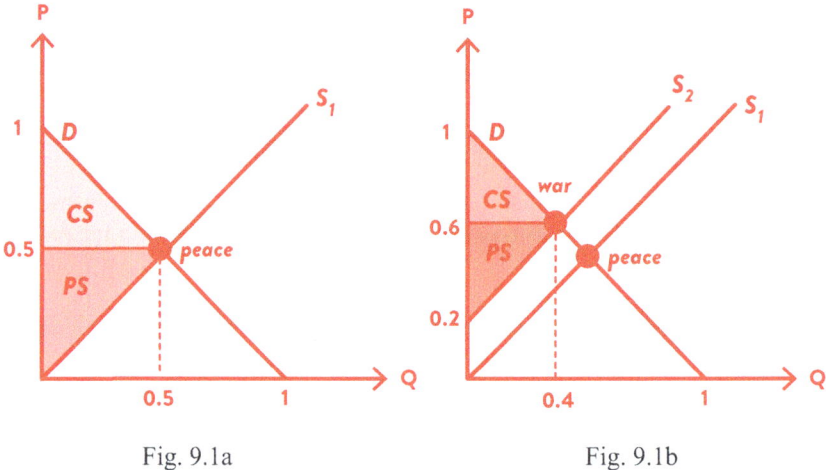

Fig. 9.1a Fig. 9.1b

The consumer surplus with peace is given by:

$$CS_{peace} = \frac{(1 - 0.5)0.5}{2} = 0.125$$

The producer surplus in this case is:

$$PS_{peace} = \frac{(0.5)0.5}{2} = 0.125$$

The total economic surplus (TS) in the peace-time market solution is thus:

$$TS_{peace} = CS_{peace} + PS_{peace} = 0.125 + 0.125 = 0.25$$

b. The supply shock ($\gamma_{war} = 0.2$) causes a new market equilibrium with $P_{war} = 0.6$ and $Q_{war} = 0.4$. This is illustrated as the point "war" in Fig. 9.1b. Consumer surplus is then:

$$CS_{war} = \frac{(1 - 0.6)0.4}{2} = 0.08$$

And producer surplus:

$$PS_{war} = \frac{(0.6 - 0.2)0.4}{2} = 0.08$$

We see that both consumer and producer surpluses have decreased due to the reduced availability of gas from Russia. The total economic surplus is therefore also lower:

$$TS_{war} = CS_{war} + PS_{war} = 0.08 + 0.08 = 0.16$$

Although the war causes a reduction in the economic surplus, there is no efficiency loss. Formally, this is because the marginal cost equals the marginal willingness to pay in the war-time equilibrium, and thus the criterion for economic efficiency is fulfilled.

A higher gas price does not mean the market is failing—the problem is the war, not the market! Intuitively, you can think of the higher price as sending a signal to consumers to reduce consumption during a time of higher production costs, which makes sense, right?

c. Just as Conrad was mistaken in thinking the high price caused by the war was a sign of economic inefficiency, he is also mistaken about the transition back to a peaceful world.

The economic surplus is obviously higher during peace than war, but the market is efficient in both cases in the sense that the market price creates the highest economic surplus.

9.2 Labour Market and Minimum Wage

a. The labour market equilibrium is given by point a in Fig. 9.2a, with wage w_a and employment L_a. When analysing the labour market, we need to think a bit differently about producer and consumer surplus because the roles are reversed compared to the product market (which has price P on the vertical axis and quantity Q on the horizontal axis).

In the labour market the producers are the firms that demand labour, while the consumers are the workers who supply labour. Producer surplus is therefore the area under the labour demand curve and above the wage, shown as area A in Fig. 9.2a. This corresponds to the operating profits generated by employing L_a workers.

Consumer surplus is the area above the labour supply curve and below the wage, area B. Since the labour supply curve is vertical here (fixed labour supply), consumer surplus is equal to the total wage income of the workers.

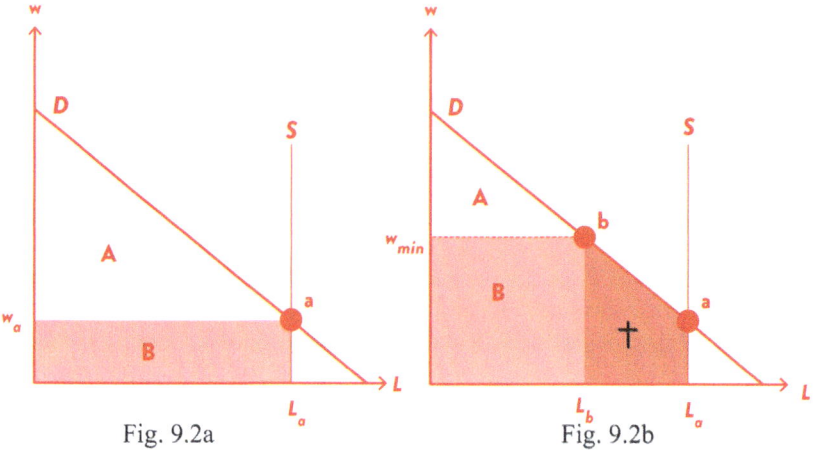

Fig. 9.2a   Fig. 9.2b

b. If the government introduces a minimum wage w_{min}, firms will adjust their labour demand to a new point b, reducing the quantity of labour demanded to L_b, as illustrated in Fig. 9.2b.

We observe that the wage increases and area B, which represents consumer surplus (for the workers), also increases. This is beneficial for the workers as a group. However, there are some losers: the producers (firms), whose surplus decreases (the triangle A showing producer surplus shrinks), and the workers who lose their jobs and become unemployed, with unemployment given by $L_a - L_b$.

In total, the minimum wage leads to a deadweight loss, represented by the area †. This deadweight loss arises because firms have a willingness to pay for labour that is higher than the wage workers would be willing to accept (which is zero here, since we are on the horizontal part of the labour supply curve). In other words, the minimum wage prevents mutually beneficial transactions from taking place.

If the government wants to increase the welfare of workers—which is a reasonable goal—a better solution would be to allow the market to determine the wage and instead impose a tax on producers (firms) and transfer payments to workers. This way, the deadweight loss caused by the minimum wage could be avoided.

9.3 The Price of Cheap Petrol

The low-price policy means that local consumers only pay the marginal cost, which implies that consumer equilibrium will be at point c, as in autarky. Total production is given by Q_b, where the marginal cost (represented by the supply curve) equals the international price. The quantity sold domestically is then Q_c, while exports become $Q_{exp} = Q_b - Q_c$.

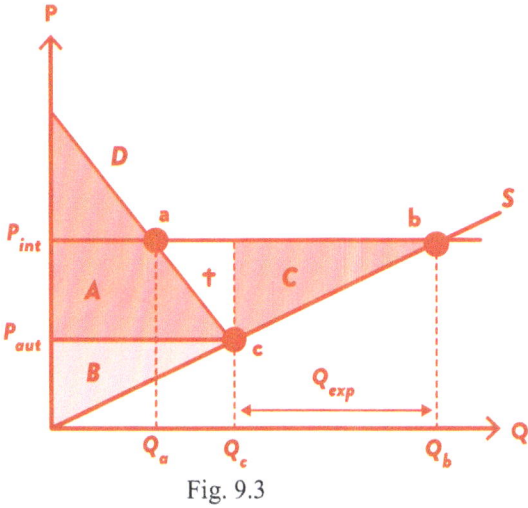

Fig. 9.3

We note that the low-price policy leads to lower exports, since a larger share of total production is consumed domestically. In a free-trade equilibrium, without a low-price policy at home, producers would sell a quantity Q_a to local consumers, while exports would be $Q_{exp} = Q_b - Q_a$.

We observe that the low-price policy creates a deadweight loss †. This arises because oil is sold domestically to customers whose willingness to pay is below the international price. While home consumers may like this policy (the consumer surplus is given by A with the low-price policy), the producers dislike it (the producer surplus shrinks: it is now B from domestic sales and C from exports), and the smaller producer surplus dominates the higher consumer surplus, which explains the efficiency cost: society may have to pay a high price for cheap petrol!

9.4 When the Invisible Hand Fails: Deadweight Loss from Pollution
a. We find the market equilibrium by combining the supply and demand functions $Q^S = -1 + 2P$ and $Q^D = 5 - P$. At equilibrium, demand equals supply, $Q^D = Q^S$ so:

$$5 - P = -1 + 2P$$

This means that the equilibrium price is $P_a = 2$. Plug this into either the demand or supply function to find the equilibrium quantity $Q_a = 3$. This is point a in Fig. 9.4 (in the textbook), shown below as Fig. 9.4 with the relevant numbers.

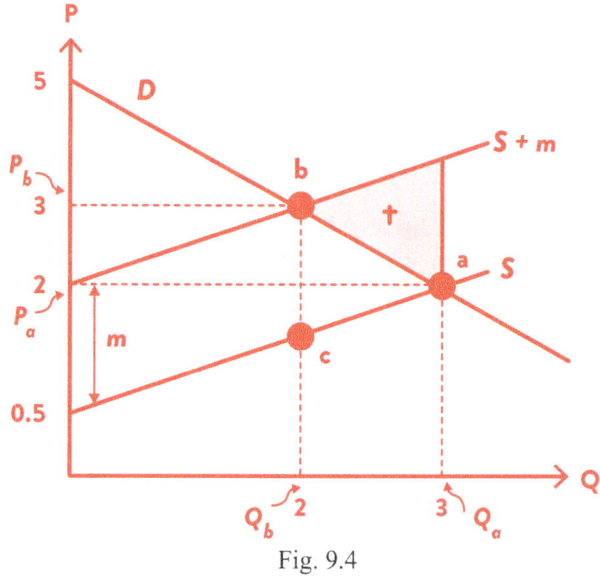

Fig. 9.4

To find the social marginal cost, start by expressing the firm's supply function $Q^S = -1 + 2P$ in inverse form (price as a function of quantity) $P = 0.5 + 0.5Q^S$. This expression represents the firm's marginal cost.

The social marginal cost $S + m$ then becomes $P = 0.5 + 0.5Q^S + m$.

Since $m = 1.5$, this simplifies to $P = 2 + 0.5Q^S$. Placing quantity on the left-hand side, we get $Q^S = -4 + 2P$. Since in equilibrium $Q^D = Q^S \equiv Q$, we can find point b in the figure by solving:

$$5 - P = -4 + 2P$$

This implies that the optimal price from the perspective of society is:

$$P_b = 3$$

Plug this into either the supply or demand function and we find the quantity in point b as:

$$Q_b = 2$$

b. The size of the deadweight loss $\dagger$ is found as the area of the shaded triangle, that is $\dagger = 0.5m(Q_a - Q_b)$. Here, the social costs are higher than the willingness to pay. Substituting the values we found above gives:

$$\dagger = 0.5m(Q_a - Q_b) = 0.5(1.5)(3 - 2) = 0.75$$

9.5 Are Imports a Threat to Innovation and Economic Efficiency?

a. We substitute the price $P_{int} = 4$ into the producers' supply function $Q^S = -2 + P$, and find that $Q^S = -2 + 4 = 2$. This is marked as point a in Fig. 9.5. Similarly, we find consumption as $Q^D = 10 - 4 = 6$, marked as point c. Imports are thus $Q^D - Q^S = 6 - 2 = 4$, which is the distance between points c and a.

b. Externalities imply that the market does not deliver the best solution for society. The economically efficient solution is found where the social marginal cost curve $S - h$ intersects the international price line $P_{int} = 4$, which means $Q^S = 4$. This is marked as point b in the figure.

When calculating the deadweight loss, we compare the market solution at point a with the ideal solution at point b, and the loss is the difference in total economic surplus between the two solutions. In the figure, the deadweight loss is given by the shaded triangle † $= 0.5(4 - 2)(4 - 2) = 2$.

To understand this, start at the ideal point b and then think of moving towards the market solution a, i.e., a reduction in production. When production is reduced by one unit, we must import this unit, and the price for that is P_{int}. However, with knowledge externalities, we see that it would be cheaper for society to produce this unit itself rather than import it, since the $S - h$ curve lies below P_{int}. Society thus loses out by reducing production, and the loss is the difference between P_{int} and the social marginal cost $S - h$. When we reach point a, the loss has become the triangle †, and this is the deadweight loss associated with the market solution (compared to the ideal solution b).

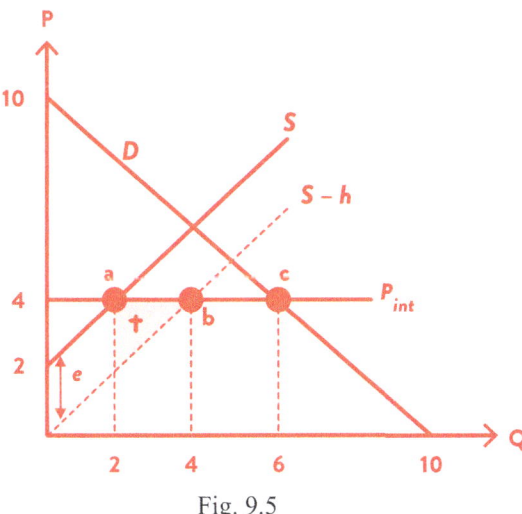

Fig. 9.5

We see that in the economically efficient solution, imports are lower than in the market solution: imports should be given by the distance between points c and b, not the distance between c and a, which is the market solution. In this way, one can say that, at least in this case, the market solution results in too much import.

More production should take place domestically, since this generates innovation and human capital with positive spillover effects for society.

Chapter 10 Economic Policy

10.1 Lower Electricity Tax, Same Bill!

a. The price that equates supply and demand is given by:

$$Q^S = Q^D \Rightarrow 1 = 1.5 - P^D$$

$$1 = 1.5 - P^D \Rightarrow P^D = 0.5$$

The equilibrium is at point a in Fig. 10.1. Without any tax, this is also the price the producer receives.

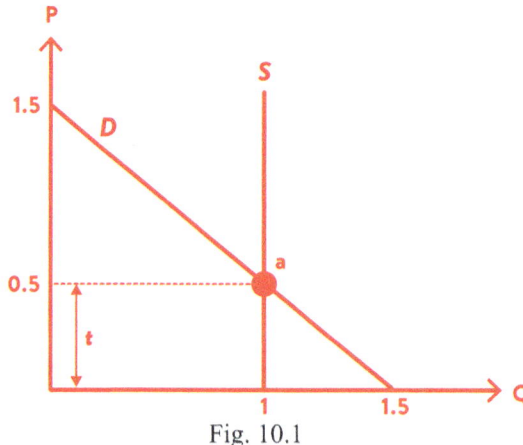

Fig. 10.1

b. We then introduce an electricity tax $t = 0.5$, which creates a wedge between the consumer price and the producer price, $P^D = P^S + t$. If the production capacity limit is binding, we have $Q^S = 1$, so that the market equilibrium is given by:

$$Q^S = Q^D \Rightarrow 1 = 1.5 - P^D$$

This can be expressed as:

$$Q^S = Q^D \Rightarrow 1 = 1.5 - \left(P^S + t\right)$$

We substitute $t = 0.5$, and get:

$$Q^S = Q^D \Rightarrow 1 = 1.5 - \left(P^S + 0.5\right) \Rightarrow P^S = 0$$

This means that the consumer price with the electricity tax becomes $P^D = P^S + t = 0.5$, the same as before. In this case, the producers bear the entire tax: the price they receive decreases one-to-one with the electricity tax, while the consumer price remains unchanged. The point is that the supply curve here is completely inelastic (i.e., vertical), which means that producers are willing to supply this quantity (given by $Q = 1$) at any price.

c. By removing the electricity tax, we effectively reverse the process in part (b), where we studied what happens when an electricity tax is introduced. When the tax is now removed, we return to a market equilibrium at point a, where $P^D = P^S = 0.5$. In other words, removing the electricity tax in this case has not helped the consumers at all; only the producers benefit from it. Just as the producers bear the full cost of the tax increase, they also receive the full benefit of removing the tax.

10.2 Subsidies, Incidence and Deadweight Loss

a. With demand $Q^D = 100 - P$ and supply $Q^S = -20 + P$, we have a market equilibrium:

$$Q^D = Q^S \Rightarrow 100 - P = -20 + P \Rightarrow P = 60$$

This gives a quantity bought and sold $Q = 40$. The equilibrium is shown as point a in Fig. 10.2a.

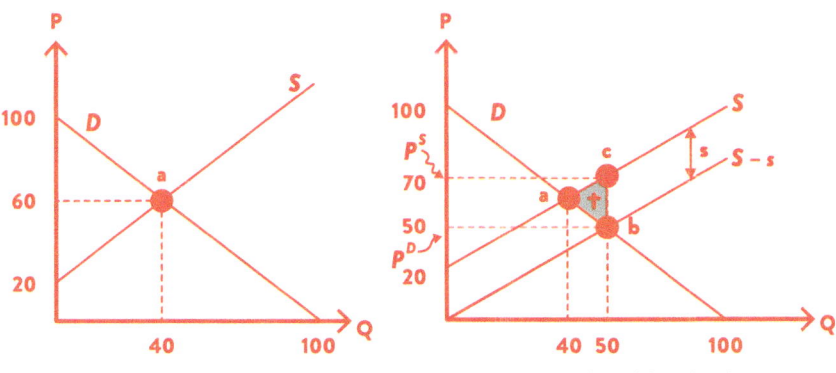

Fig. 10.2a Without subsidy.   Fig. 10.2b With subsidy

b. The Producer Party has managed to introduce a production subsidy s, which creates a wedge between the consumer price and the producer price, so that $P^D + s = P^S$.

The demand function can now be written as $Q^D = 100 - P^D$, while the supply function becomes $Q^S = -20 + P^S$. In equilibrium, demand must still equal supply, which means:

$$Q^D = Q^S \Rightarrow 100 - P^D = -20 + P^S$$

Using $P^D + s = P^S$ in the above expression, we get:

$$100 - P^D = -20 + \left(P^D + s\right)$$

Since $s = 20$ this becomes:

$$100 - P^D = -20 + \left(P^D + 20\right)$$

Which simplifies to:

$$P^D = 50$$

At this price, consumers will choose the equilibrium at point b in Fig. 10.2b, with $Q^D = 50$. Since $P^D + s = P^S$, we can find the producer price as:

$$P^S = 50 + 20 = 70$$

c. Notice that the production subsidy has led to a lower consumer price. This is the subsidy version of tax incidence: even though the subsidy goes to producers, consumers also benefit. In this example, the consumer price falls just as much as the producer price rises, so the two sides share the subsidy equally. The Producer Party will probably be somewhat dissatisfied with this outcome!

What about the Efficiency Party? We observe that the policy creates a deadweight loss as the subsidy causes production to increase, and marginal cost becomes higher than the marginal willingness to pay, shown by the gap between the supply curve S and the demand curve D.

The size of the deadweight loss is given by the triangle abc:

$$\dagger = 0.5 \times (70 - 50) \times (50 - 40) = 100$$

The Efficiency Party would not support this policy!

10.3 The Environmentalists

a. With constant marginal cost, the price in a perfectly competitive market is also constant, and demand determines the quantity produced. Without any government intervention, the price will be $P = MC = 50$, and demand, and thus production, will be $Q = 100 - P = 100 - 50 = 50$. This is illustrated by point a in Fig. 10.3a, with $Q_a = 50$ and $P_a = 50$.

The exercise states that firms pollute, and this costs society $e = 10$ per unit produced. Firms do not take its pollution into account when deciding their output: they only consider the private marginal cost of 50, not the social marginal cost, which is $MC + e = 50 + 10 = 60$. The optimal allocation, where marginal willing-ness to pay equals the social marginal cost, is at $Q^* = 40$ and $P^* = 60$, marked by point b in Fig. 10.3a. The deadweight loss, i.e., the area where willingness to pay is below the social cost, is given by:

$$\dagger = 0.5(P^* - P_a)(Q_a - Q^*) = 0.5(60 - 50)(50 - 40) = 50$$

The economic surplus under the market solution consists of consumer surplus minus environmental costs (note that with horizontal marginal cost there is no producer surplus):

$$TS_a = CS_a - eQ_a = 0.5(100 - 50)50 - 10(50) = 750$$

b. As discussed in the textbook, the optimal environmental tax is equal to the marginal environmental cost per unit: $t^* = 10$. This means the firms will take the full social cost into account when making its production decisions. The resulting production will be $Q^* = 40$, and the economic surplus now includes not just the consumer surplus but also the tax revenue $T = tQ$, minus the environmental costs. The economic surplus becomes:

$$TS^* = CS^* - eQ^* + t^*Q^* = 0.5(100 - 60)40 - 10(40) + 10(40) = 800$$

We see that $TS^* - TS_a = \dagger$, as expected.

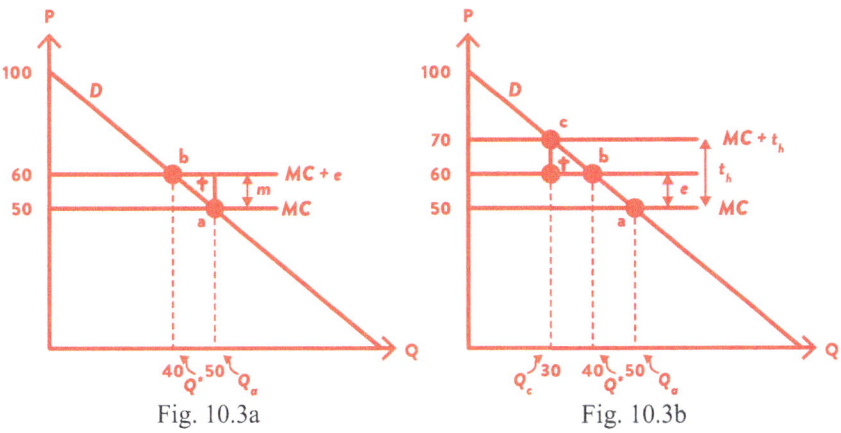

Fig. 10.3a   Fig. 10.3b

c. The Green Party proposes a higher environmental tax, $t_{high} = 20$. This leads to a price of $MC + t_{high} = 70$. As shown in Fig. 10.3b, consumer equilibrium now is at point c. Compared to the optimal solution in point b, this gives rise to an efficiency loss because willingness to pay exceeds the social cost. With $Q_c = 30$ and $P_c = 70$, the efficiency loss from the high tax is:

$$\dagger = 0.5(P_c - P^*)(Q^* - Q_c) = 0.5(70 - 60)(40 - 30) = 50$$

We see that the efficiency loss from the high tax suggested by the Green Party is just as large as without any tax at all. The social surplus with the high tax is therefore the same as the market solution found in part (a):

$$TS_c = CS_c - eQ_c + t_{high}Q_c = 0.5(100 - 70)30 - 10(30) + 20(30) = 750$$

Without an environmental tax, production is too high. With the high tax, however, production is too low. A tax rate higher than t_{high} would clearly lead to an even larger loss in economic surplus.

This illustrates an important point: pollution means the activity should be limited, but the level of the tax must be balanced against other considerations. For instance, banning a polluting activity is usually not optimal. The optimal policy involves a trade-off between different interests—environmental concerns are important, but they must be weighed against the interests of consumers (and producers).

10.4 Trade Policy for Innovation?
This exercise follows up on Exercise 9.5, but here we take a closer look at the issue and discuss the best policy for achieving the socially optimal outcome. We also calculate consumer and producer surplus to examine distributional effects.
a. Autarky leads to an equilibrium at point a in Fig. 10.4. In this case, the consumer surplus is:

$$CS_{aut} = \frac{(10 - 6)4}{2} = 8$$

The producer surplus becomes

$$PS_{aut} = \frac{(6 - 2)4}{2} = 8$$

The value of the human capital created in society is:

$$H_{aut} = hQ_{aut}^S = 2(4) = 8$$

Total economic surplus in autarky is therefore:

$$TS_{aut} = CS_{aut} + PS_{aut} + H_{aut} = 8 + 8 + 8 = 24$$

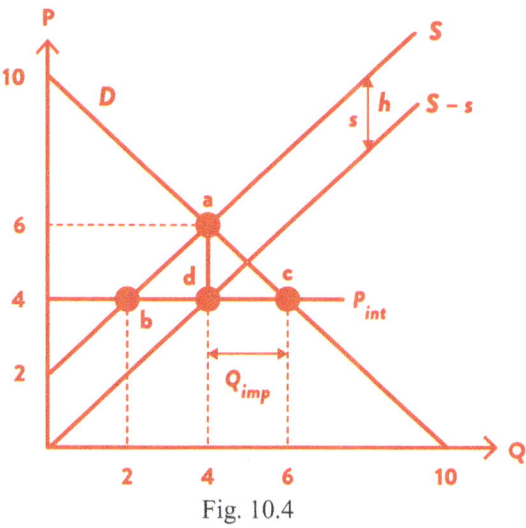

Fig. 10.4

b. We know from the discussion of environmental taxes—so-called Pigouvian taxes—that these can be used to internalise an externality and achieve a socially optimal outcome (see Sect. 10.7 in the textbook). In this case, with a positive externality, the optimal policy is a production subsidy (s) equal to the size of the external effect $s = h = 2$.

This means the supply curve shifts to $S - s$. With this subsidy, producers choose to produce at point d, where $S - s$ intersects the international price P_{int}.

We see that production in this example is the same as under autarky $Q^S = 4$ while consumers choose point c, where $Q^D = 6$, and the difference between consumption and production is imports $Q_{imp} = 6 - 4 = 2$.

What is the total economic surplus in this scenario? The consumer surplus, with consumers choosing point c, is:

$$CS^*_{int} = \frac{(10 - 4)6}{2} = 18$$

The producer surplus (including the subsidy), given their choice of point d, is:

$$PS^*_{int} = \frac{(4)4}{2} = 8$$

The value of the human capital generated is:

$$H_{int} = hQ^S_{int} = 2(4) = 8$$

The subsidy costs:

$$S = sQ^S = 2(4) = 8$$

Total economic surplus is therefore:

$$TS^*_{int} = CS^*_{int} + PS^*_{int} + H_{int} - S = 18 + 8 + 8 - 8 = 26$$

We therefore see that $TS^*_{int} > TS_{aut}$. A production subsidy combined with allowing international trade is therefore better than autarky.

The difference is $(26 - 24) = 2$, and in the figure this is illustrated by the shaded triangle acd.

c. Why is a production subsidy better than autarky? Autarky and the production subsidy result in the same increase in production and thus the same positive effect on innovation, but the difference lies in the fact that autarky restricts consumption.

The general lesson is that economic policy should be targeted. If there is a positive externality linked to production, the response should be to provide a targeted reward to production itself. An import ban does provide such a reward, but it also comes with side effects.

Economic policy is like medicine: the treatment should match the need. For example, broad-spectrum antibiotics should be used with caution, since they also eliminate good bacteria!

Similarly, autarky may support innovation, but such a policy imposes negative side effects on consumers.

10.5 Odd–Even Driving vs. Road Tolls

a. Odd–even driving means that only half of all cars are allowed on the roads, regardless of the driver's willingness to pay.

Without any driving restrictions, demand is given by $Q = 1 - P$, but under odd–even driving, demand becomes $Q' = 0.5(1 - P)$. This means that the new demand curve (shown as the demand curve *Demand'* in Fig. 10.5a) intersects the horizontal axis at $P = 0 \Rightarrow Q' = 0.5$ while the choke price remains the same $Q' = 0 \Rightarrow P^{choke}_D = 1$ (since the policy does not affect individual willingness to pay, it only limits how many people are allowed through).

Since the price to drive into the city is zero, everyone who is allowed to drive $(Q = 0.5)$ will do so, and the consumers choose point b. The consumer surplus, which in this case is also the total economic surplus, is therefore given by the shaded area A in Fig. 10.5a.

$$TS_a = CS_a = \frac{1(0.5)}{2} = 0.25$$

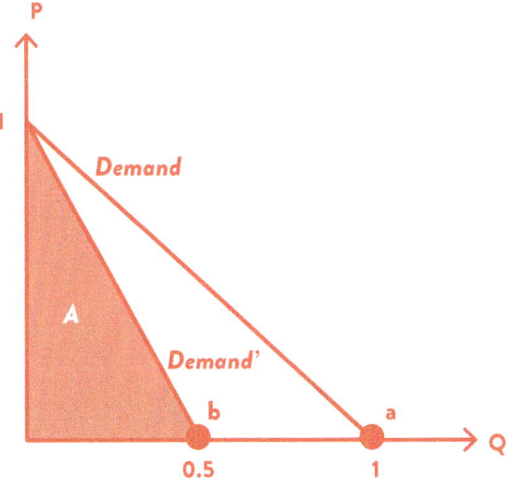

Fig. 10.5a Odd-even driving

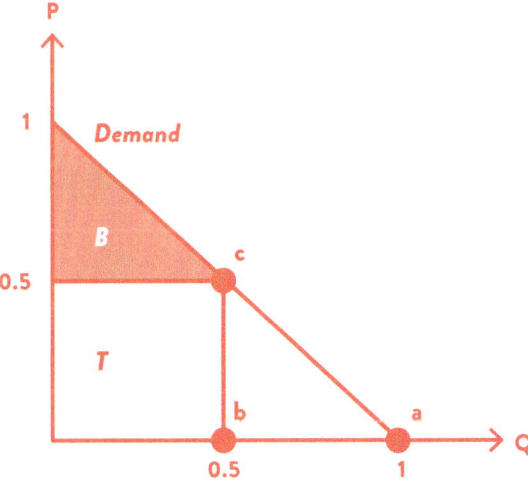

Fig. 10.5b Congestion pricing

b. With a road toll of $P = 0.5, Q = 0.5$ cars will want to enter the city, thus achieving the same reduction in traffic as the odd-even policy. This is illustrated in Fig. 10.5b, with the equilibrium at point c, consumer surplus B and the municipality's toll revenue represented by area T.

The size of the consumer surplus is:

$$CS_b = 0.5(1 - P)Q = 0.5(0.5)0.5 = 0.125$$

While the municipality's toll revenue is:

$$T = PQ = 0.5(0.5) = 0.25$$

The total economic surplus is now:

$$TS_b = CS_b + T = 0.125 + 0.25 = 0.375$$

c. The total economic surplus with the road toll is higher than with odd–even driving: $TS_b = 0.375 > TS_a = 0.25$.

In Fig. 10.5c, the difference in total surplus is given by the area C + D, which in turn equals the surplus B + T (from Fig. 10.5b, with the road toll) minus the area A (from Fig. 10.5a, with odd-even driving).

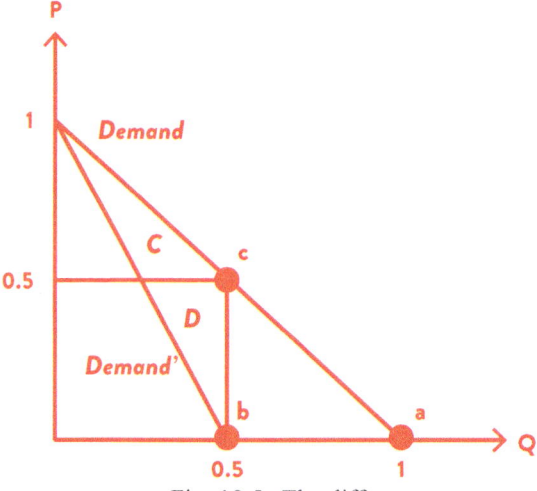

Fig. 10.5c The difference

How can we explain this? After all, the number of cars is the same in both cases!

The answer lies in *who* gets to drive. From the perspective of economic efficiency, odd–even driving leaves out and lets in the wrong people!

Imagine a switch from the toll to odd–even driving. The area C is the lost consumer surplus for those with a high willingness to pay who are no longer allowed to drive into the city under odd–even rules. The area D represents the loss from letting in consumers with a willingness to pay lower than the road toll.

In sum, odd–even driving means that many people with a high willingness to pay are not allowed to drive, while many with a low willingness to pay enter the city. As we've already seen from the analysis above, $TS_b - TS_a = 0.375 - 0.25 = 0.125$ so the area C + D equals 0.125.

In the textbook, we discuss environmental taxes and tradable CO_2 quotas. The road toll in the current exercise acts like an environmental tax: it limits the extent of an undesirable activity. And as you've seen, tolls are a more efficient way to manage traffic into the city centre than a random allocation of permits through odd–even driving.

Chapter 11 Monopoly

11.1 "Green is good"—But at What Price?

a. The price of Very Vegan cannot be optimal from a profit-maximizing perspective. It implies *inelastic demand* (i.e., $\varepsilon < 1$), since the percentage reduction in quantity demanded is smaller than the percentage increase in price. But we know that a profit-maximising monopolist will always choose a price and quantity combination where demand is *elastic* (see Fig. 11.2 in the textbook). The foreign publisher should therefore increase the price to increase profits.

b. For the linear demand function $Q = \alpha - \beta P$, which can be written in inverse form as $P = (\alpha - Q)/\beta$, the price elasticity of demand is:

$$\varepsilon = -\frac{\partial Q}{\partial P}\frac{P}{Q} = \beta\frac{\alpha - Q}{\beta Q} = \frac{\alpha - Q}{Q}$$

According to the market study, $\varepsilon = 0.5$. We are also told that $Q = 2/3$. Substituting into the elasticity formula to find α:

$$\varepsilon = \frac{\alpha - Q}{Q} \Rightarrow 0.5 = \frac{\alpha - \frac{2}{3}}{\frac{2}{3}} \Rightarrow \frac{2}{6} = \alpha - \frac{2}{3} \Rightarrow \alpha = 1$$

We use $\alpha = 1$ along with the information given in the problem, $P = 1/3$ and $Q = 2/3$ and substitute into the demand function to find β:

$$Q = \alpha - \beta P \Rightarrow \frac{2}{3} = 1 - \beta\frac{1}{3} \Rightarrow \beta = 1$$

With $\alpha = \beta = 1$ the demand function can be written as:

$$Q = 1 - P$$

This is illustrated in Fig. 11.1a.

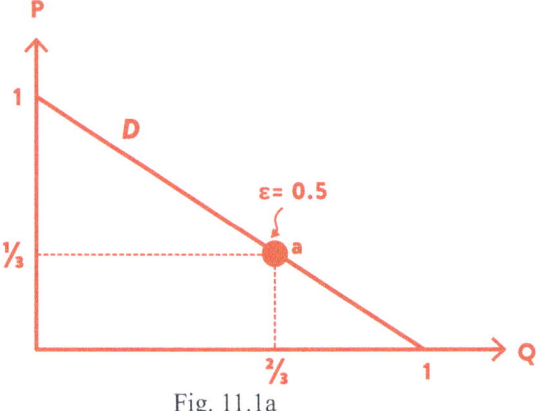

Fig. 11.1a

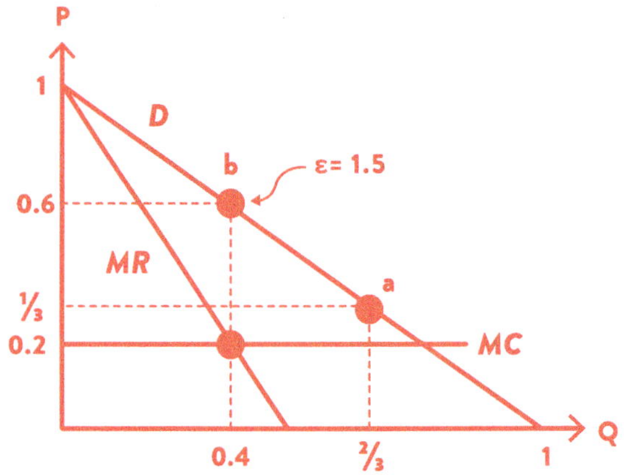

Fig. 11.1b

c. With a constant marginal cost ($c = 0.2$) we can use the mid-price rule to find the monopoly price:

$$P_{Mon} = \frac{1}{2}\left(\frac{\alpha}{\beta} + c\right) = 0.5(1 + 0.2) = 0.6$$

Alternatively, we could have started from the optimality condition $MR = MC$, where $MR = 1 - 2Q$, meaning that the marginal revenue curve falls twice as fast as the demand curve. We then have $MR = MC \Rightarrow 1 - 2Q = 0.2 \Rightarrow Q = 0.4$, which implies that $P = 0.6$. The price should therefore increase from 1/3 to 0.6, from point a to point b in Fig. 11.1b.

d. At the optimum, with $P = 0.6, Q = 0.4$, we have $\varepsilon = \frac{\alpha - Q}{Q} = \frac{1 - 0.4}{0.4} = 1.5$, as illustrated in Fig. 11.1b. We see that the demand is elastic at this point, in line with the theory for a profit-maximising monopolist.

11.2 The Monopolist on the Mountain

a. The Lerner index is the markup over marginal cost, expressed as a share of the price:

$$\frac{P - MC}{P} = \frac{6 - 4}{6} = \frac{1}{3} \quad \text{The kiosk owner's claim}$$

From Chapter 11.4 in the textbook, we know that the Lerner index for a profit-maximising monopolist is:

$$\frac{P - MC}{P} = \frac{1}{\varepsilon}$$

The information from the kiosk monopolist, with a markup of 1/3, thus implies that $\varepsilon = 3$, which is a high price elasticity of demand.

b. Anna's survey shows that a 10% increase in the price of soda would lead to a 15% reduction in demand. This means a price elasticity of demand $\varepsilon = 1.5$, which is much lower than the elasticity we found in part (a) based on the kiosk owner's information ($\varepsilon = 3$ implies that a 10% price increase would reduce demand by 30%).

We use this to find the marginal cost based on the Lerner index:

$$\frac{P - MC}{P} = \frac{1}{\varepsilon} \Rightarrow \frac{6 - MC}{6} = \frac{1}{1.5} = \frac{2}{3} \quad \text{Anna's claim}$$

Anna's estimate of the markup is thus twice as large as the kiosk owner's claim, implying a lower marginal cost:

$$\frac{6 - MC}{6} = \frac{2}{3} \Rightarrow MC = 2$$

That is, Anna suggests it costs the kiosk only half as much as the owner claimed to buy and transport the soda to the top of the mountain, 2 euros per bottle, not 4 euros.

c. Since the kiosk has constant marginal costs, we can use the midprice rule, which states that:

$$P_{Mon} = \frac{1}{2}\left(\frac{\alpha}{\beta} + c\right)$$

We substitute the observed price of a soda, $P_{Mon} = 6$, and use the kiosk owner's claimed marginal cost $c = 4$:

$$6 = \frac{1}{2}\left(\frac{\alpha}{\beta} + 4\right) \Rightarrow 8 = \frac{\alpha}{\beta} \Rightarrow \alpha = 8\beta$$

We use $\alpha = 8\beta$ and the given information about the price $P = 6$ and that the kiosk sells four bottles per day, $Q = 4$, in the demand function to find β:

$$Q = \alpha - \beta P \Rightarrow 4 = 8\beta - \beta 6 \Rightarrow \beta = 2$$

Since $\alpha = 8\beta$, we then know that $\alpha = 8(2) = 16$ so the demand curve consistent with the owner's information is:

$$Q = \alpha - \beta P = 16 - 2P$$

The choke price for consumers here is:

$$P_D^{choke} = \frac{\alpha}{\beta} = 8$$

That is, no hiker is willing to pay more than 8 euros for a bottle of soda, according to the information from the kiosk.

The market outcome described by the kiosk owner is then as shown in Fig. 11.2a.

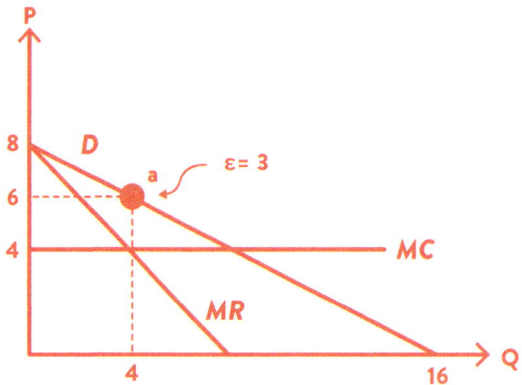

Fig. 11.2a Kiosk owner's claim

Anna's estimate is different. We substitute the observed price of a soda, $P_{Mon} = 6$, and use the marginal cost she found, $c = 2$:

$$6 = \frac{1}{2}\left(\frac{\alpha}{\beta} + 2\right) \Rightarrow 10 = \frac{\alpha}{\beta} \Rightarrow \alpha = 10\beta$$

The rest of the procedure is the same as when we found the demand parameters based on the kiosk owner's claim. Using Anna's figures in the demand function, we get:

$$Q = \alpha - \beta P \Rightarrow 4 = 10\beta - \beta 6 \Rightarrow \beta = 1$$

Since $\alpha = 10\beta$, we know that $\alpha = 10$ so that the demand function according to Anna is:

$$Q = \alpha - \beta P = 10 - P$$

Consumer choke price is then:

$$P_D^{choke} = \frac{\alpha}{\beta} = 10$$

That is, a far higher choke price than what the kiosk owner claimed. The market equilibrium according to Anna's calculations is shown in Fig. 11.2b.

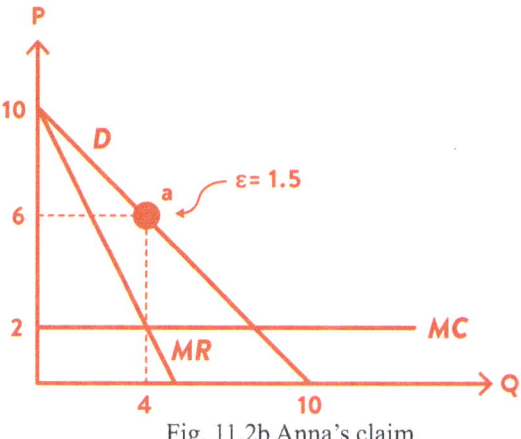

Fig. 11.2b Anna's claim

11.3 One Monopolist, Two Markets, Two Cost Scenarios

With constant marginal costs, $c = 1/3$, and we can therefore use the midprice rule:

$$P_{Mon} = \frac{1}{2}\left(P_D^{choke} + c\right)$$

In market A the choke price is:

$$P_A^{choke} = \frac{\alpha}{\beta} = \frac{2}{2} = 1$$

And similarly for market B:

$$P_B^{choke} = \frac{\alpha}{\beta} = \frac{1}{1} = 1$$

With an identical choke price in the two markets $P_D^{choke} = 1$ and $c = 1/3$, the optimal price in the two markets is therefore the same:

$$P_A^{Mon} = P_B^{Mon} = \frac{1}{2}\left(1 + \frac{1}{3}\right) = \frac{2}{3}$$

Plugging into the respective demand functions we find quantities as:

$$Q_A^{Mon} = 2 - 2P = 2 - \frac{4}{3} = \frac{2}{3}$$

$$Q_B^{Mon} = 1 - P = 1 - \frac{2}{3} = \frac{1}{3}$$

The equilibria are illustrated in Fig. 11.3a for market A and 11.3b for market B.

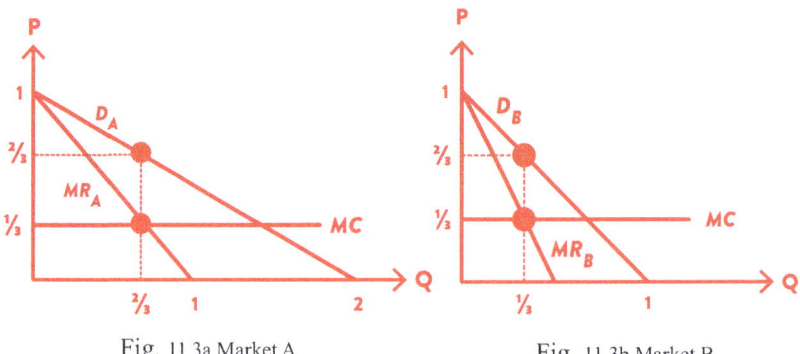

Fig. 11.3a Market A   Fig. 11.3b Market B

If total costs are $TC = \frac{1}{2}Q^2$, the marginal cost is $MC = Q$. Since the marginal cost is not constant here, we cannot use the midprice rule. Instead, we use the optimality condition that marginal revenue equals marginal cost and start by finding the marginal revenue.

Generally, we can express demand $Q = \alpha - \beta P$ in inverse form as:

$$P = \frac{\alpha - Q}{\beta}$$

Revenue, $R = PQ$, is then:

$$R = \left(\frac{\alpha - Q}{\beta}\right)Q$$

And the marginal revenue can be found as:

$$MR = \frac{\partial R}{\partial Q} = \frac{\alpha - 2Q}{\beta}$$

In profit-maximising optimum we have:

$$MR = MC \Rightarrow \frac{\alpha - 2Q}{\beta} = Q$$

$$Q_{Mon} = \frac{\alpha}{2 + \beta}$$

Plugging into the inverse demand function we find the price as:

$$P = \frac{\alpha - Q}{\beta} = \frac{\alpha - \left(\frac{\alpha}{2+\beta}\right)}{\beta} = \frac{(2 + \beta)\alpha - \alpha}{\beta(2 + \beta)} = \frac{\alpha(1 + \beta)}{\beta(2 + \beta)}$$

We substitute the given values for demand in market A, $\alpha = \beta = 2$:

$$P_A^{Mon} = \frac{2(1 + 2)}{2(2 + 2)} = \frac{3}{4}$$

Similarly for market B, where $\alpha = \beta = 1$:

$$P_B^{Mon} = \frac{1(1 + 1)}{1(2 + 1)} = \frac{2}{3}$$

Using these prices in the demand functions we find quantities as:

$$Q_A^{Mon} = 2 - 2P = 2 - \frac{6}{4} = \frac{1}{2}$$

$$Q_B^{Mon} = 1 - P = 1 - \frac{2}{3} = \frac{1}{3}$$

Figures 11.3c and d illustrate the monopolist's choice in the two markets when marginal costs are increasing.

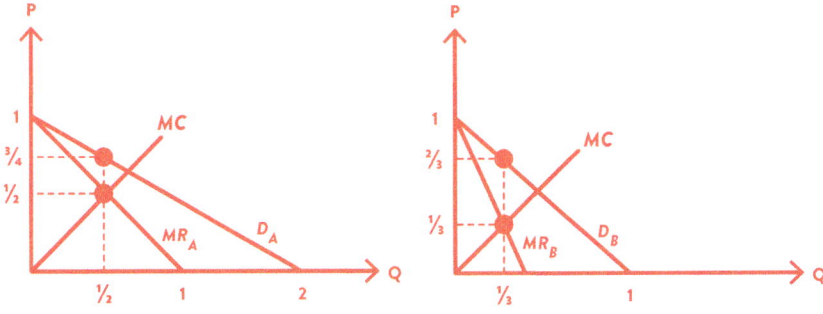

Fig. 11.3c Market A   Fig. 11.3d Market B

We see that in market A, the price is lower and the quantity higher in the scenario with constant marginal costs compared to the current scenario with increasing marginal costs.

In market B, however, the price and quantity produced are the same in both cost scenarios. This is because the marginal cost crosses the marginal revenue at the same point in both cases.

11.4 Efficiency Loss Under Monopoly

We know from the previous exercise that the monopoly price and quantity in both cost scenarios are $P^{Mon} = 2/3$ and $Q^{Mon} = 1/3$.

To find the efficiency loss, we must contrast the monopoly solution with that under perfect competition, given by price equal marginal cost.

With $c = 1/3$, (Scenario A) this implies $P_A^{Comp} = 1/3$ which plugged into the demand function implies $Q_A^{Comp} = 2/3$, as indicated by point c in Fig. 11.4a. The size of the deadweight loss in this cost scenario is:

$$\dagger_A = \frac{1}{2}\left(\frac{2}{3} - \frac{1}{3}\right)\left(\frac{2}{3} - \frac{1}{3}\right) = \frac{1}{18}$$

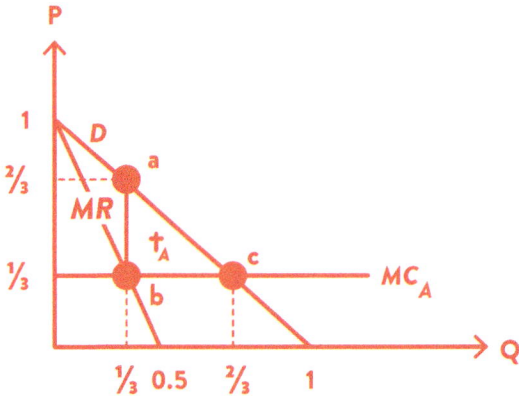

Fig. 11.4a Scenario A

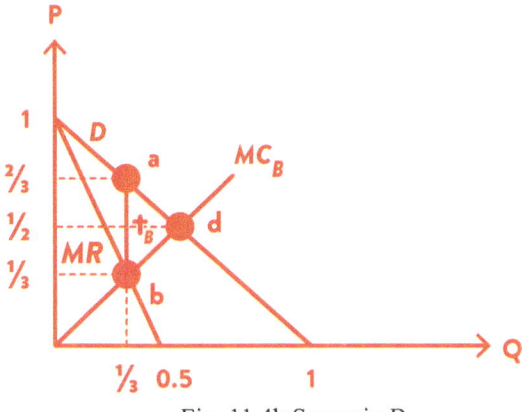

Fig. 11.4b Scenario B

With increasing marginal cost, that is, Scenario B, we have $MC = Q$, and the market outcome with competition implies $P = MC \Rightarrow P = Q$, which plugged into the demand function results in:

$$Q = 1 - P = 1 - Q \Rightarrow 2Q = 1 \Rightarrow Q_B^{Comp} = \frac{1}{2}$$

Given that $P = MC \Rightarrow P = Q$, the competitive price here will be:

$$P_B^{Comp} = \frac{1}{2}$$

The perfectly competitive equilibrium is marked as point d in Fig. 11.4b. The efficiency loss is now:

$$\dagger_B = \frac{1}{2}\left(\frac{2}{3} - \frac{1}{3}\right)\left(\frac{1}{2} - \frac{1}{3}\right) = \frac{1}{36}$$

c. The efficiency loss with increasing marginal costs is therefore only half of what we found when marginal costs are constant: $\dagger_B = 1/36$, while $\dagger_A = 1/18$. This is because monopoly leads to a larger reduction in the quantity produced in Scenario A than in Scenario B: $Q_A^{Comp} - Q_A^{Mon} = 2/3 - 1/3 = 1/3$ compared to $Q_B^{Comp} - Q_B^{Mon} = 1/2 - 1/3 = 1/6$. And with a larger deviation from the ideal point, that is, perfect competition, we have a larger efficiency loss.

11.5 Natural Monopoly and Economic Surplus
a. A fixed cost and constant marginal costs mean that the average total cost declines. This is therefore a natural monopoly.

Since the monopolist has constant marginal costs, we can use the midprice rule to find the monopoly price:

$$P_{Mon} = \frac{1}{2}\left(P_D^{choke} + c\right)$$

With the demand function $Q = 10 - P$, we know that the choke price is $P_D^{choke} = 10$, and with marginal cost $c = 0$, we get:

$$P_{Mon} = \frac{1}{2}(10 + 0) = 5$$

An alternative approach to find the monopolist's optimum is to use the first-order condition $MR = MC$. We find marginal revenue as:

$$MR = \frac{\partial R}{\partial Q} = P + Q\frac{\partial P}{\partial Q} = 10 - Q + Q(-1) = 10 - 2Q$$

Since marginal cost is zero, we have:

$$MR = MC \Rightarrow 10 - 2Q = 0 \Rightarrow Q = 5$$

In other words, $Q_{Mon} = 5$, and inserting into the demand function gives $P_{Mon} = 5$.

That is, the same result as with the midprice rule. The monopolist's choice is shown as point m in Fig. 11.5a. The producer surplus (which is the same as operating profits) is $PQ = 25$, while the profis are $\pi = PQ - TC = 25 - 10 = 15$, and the total economic surplus is $TS = CS + PS = 0.5(10 - 5)5 + 5(5) = 12.5 + 25 = 37.5$.

b. The socially optimal solution is where price equals marginal cost. With marginal cost equal to zero, this implies that the price should also be zero, which gives $Q = 10$.

This is shown as point f in Fig. 11.5a, giving $TS = CS + PS = 0.510(10) + 0 = 50$.

The welfare loss from the monopoly outcome is therefore $\dagger = 12.5$, as shown by the shaded triangle in Fig. 11.5a.

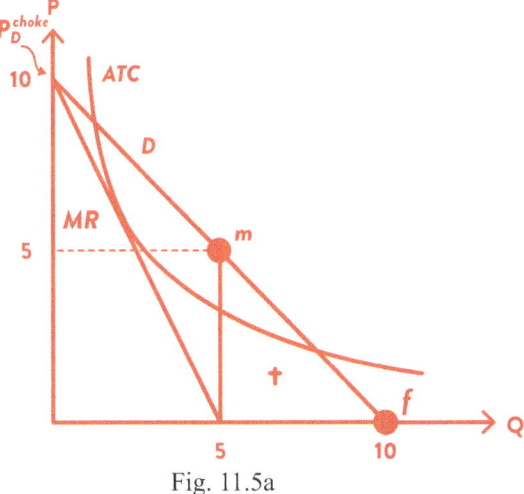

Fig. 11.5a

Note also that fixed costs are not included in the calculation of total economic surplus: producer surplus is defined as operating profits. Why? Because fixed costs are, in the short run, exactly that—fixed, and the opportunity cost is therefore zero (since nothing can be done about fixed costs in the short run).

The private economic loss, which in this case equals the fixed costs of 10, can be covered by a general tax (ideally a lump sum tax on consumers, which is simply a transfer from consumers to the state, and thus does not create any efficiency loss).

c. If the government takes over operations with a requirement of zero profits, the outcome will be as follows:

$$\pi = 0 \Rightarrow PQ = TC$$

This can be written as:

$$P = \frac{TC}{Q} - ATC$$

That is, price equals average total cost. Using the inverse demand function, $P = 10 - Q$, and the fact that ATC is simply the fixed costs (which here is 10) divided by the quantity Q, we get:

$$P = ATC \Rightarrow 10 - Q = \frac{10}{Q}$$

This can be written as:

$$Q - 10 + \frac{10}{Q} = 0$$

We multiply both sides of the equation by Q and get:

$$Q^2 - 10Q + 10 = 0$$

This is a quadratic equation of the form $ax^2 + bx + c = 0$, and we can use the quadratic formula to solve it. In our case, $a = 1, b = -10, c = 10$, and we find that:

$$Q = \frac{5 - \sqrt{15} \approx 1.13}{5 + \sqrt{15} \approx 8.87}$$

In Fig. 11.5b, the low and high values are given by the two points where the demand curve intersects the ATC curve—at $Q \approx 1.13$ and $Q \approx 8.87$. The firm should choose the lowest price, and thus the highest quantity, which means the price is

$$P \approx 1.13$$

This is shown as point g in Fig. 11.5b.

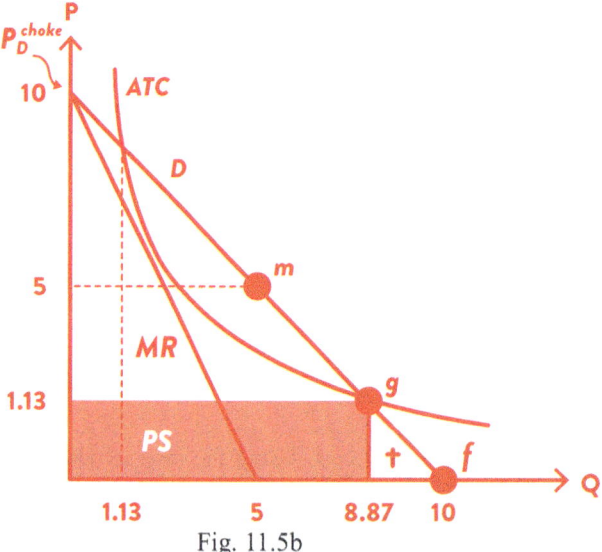

Fig. 11.5b

Compared to the solution in part (b), we see that the operating profits, which are the same as producer surplus (PS), have increased and are represented by the shaded rectangle in the figure, with an area approximately equal to 10, meaning that profits are zero. However, this area has been taken from consumers, who also lose the triangle †. It is this lost triangle, with an area approximately equal to $0.5(10 - 8.87)1.13 = 0.64$, that represents the deadweight loss of this solution compared to the situation with price equal to marginal cost.

Chapter 12 Oligopoly

12.1 Is a Merger Profitable?

a. From Math Box 12.1 (make sure you understand the method!), we see that with $Q = 1 - P$ and $c = 0$, the quantity produced in the Cournot equilibrium for each firm $i = A, B$ is:

$$Q_i^{Duo} = \frac{1}{3}$$

Since the inverse demand is $P = 1 - Q$ and $Q = Q_A^{Duo} + Q_B^{Duo} = 2/3$, we get $P_{Duo} = 1/3$ and profits (which here equals revenue, since costs are zero) for each firm is:

$$\pi_i^{Duo} = P_{Duo} Q_i^{Duo} = \frac{1}{9}$$

When A and B merge, they form the monopoly AB. You can think of the merger as a marriage, where the "couple" AB acts as a single decision-maker and splits the monopoly profits equally.

To find the monopoly profits, you can refer to Math Box 11.1, or use the midpoint rule, which in this case (with $c = 0$ and choke price $= 1$) implies $P_{Mon} = 0.5$ and with quantity:

$$Q_{AB}^{Mon} = 0.5$$

So the profits (and revenue) are:

$$\pi_{AB}^{Mon} = P_{Mon} Q_{AB}^{Mon} = 0.25$$

Each of the merging firms then receives half of this, which is greater than what they made as competitors: $0.5\pi_{AB}^{Mon} = 1/8 > \pi_i^{Duo} = 1/9$.

Since $1/8 = 0.125$ is greater than $1/9 \approx 0.111$, the merger is profitable for both firms.

b. With a triopoly, we have $Q = Q_A + Q_B + Q_C$. Let's look at Firm A. Since $c = 0$, the profits can be written as:

$$\pi_A = (1 - Q_A - Q_B - Q_C)Q_A$$

Profit maximisation gives:

$$\frac{\partial \pi_A}{\partial Q_A} = 1 - 2Q_A - Q_B - Q_C = 0$$

From this, we can find A's reaction function as:

$$R_A(Q_B, Q_C) = Q_A = \frac{1 - Q_B - Q_C}{2}$$

Since the three firms are identical, in equilibrium we must have $Q_A = Q_B = Q_C = Q_i$. The reaction functions can therefore be written as:

$$Q_i = \frac{1 - Q_i - Q_i}{2}$$

Which implies that:

$$Q_i^{Trio} = 0.25$$

Total quantity in triopoly is then:

$$Q_{Trio} = Q_A + Q_B + Q_C = 0.75$$

And the price:

$$P_{Trio} = 1 - 0.75 = 0.25$$

Each firm's profits in triopoly is therefore:

$$\pi_i^{Trio} = P_{Trio}Q_i^{Trio} = (0.25)0.25 = \frac{1}{16}$$

Now firms A and B merge, forming the merged unit AB. This leaves a duopoly between AB and C, where quantity, price, and profits for the duopolists are exactly as you calculated in part (a).

Because competition is less intense in a duopoly than in a triopoly, profits rise from $\pi_i^{Trio} = 1/16$ to $\pi_i^{Duo} = 1/9$. The merged unit AB shares the duopoly profits equally, so each gets $0.5\pi_{AB}^{Duo} = 0.5(1/9) = 1/18$.

However, their shared post-merger profits are less than what each earned separately before the merger, in the triopoly, where $\pi_i^{Trio} = 1/16$.

Thus, the "marriage" between A and B is not profitable together they earn less than they did apart. The winner from the merger is firm C, the firm outside the merger! This is called profit shifting: the merging parties reduce their output to push the price up, and the firm outside takes the opportunity to increase its output (from $Q_i^{Trio} = 0.25$ to $Q_i^{Duo} = 1/3$) thereby shifting profits from the merging parties to itself.

12.2 Foreign Ownership, Competition and Economic Efficiency

a. The monopolist's optimal quantity is found where marginal revenue equals marginal cost. The revenue is $R = PQ$ and with the inverse demand function $P = 1 - Q$, revenue becomes $R = Q(1 - Q)$, and marginal revenue becomes:

$$MR = \frac{\partial R}{\partial Q} = 1 - 2Q \quad \text{Marginal revenue}$$

With zero marginal cost, the optimum condition $MR = MC$ implies $MR = 0$, which gives us quantity:

$$Q_A^{Mon} = \frac{1}{2}$$

And price:

$$P_A^{Mon} = \frac{1}{2}$$

Note that we could alternatively have found the monopoly price using the midpoint rule. The monopolist's choice is given by point a i Fig. 12.2a. This gives the consumer surplus (area A):

$$CS_{Mon} = \frac{\left(P_D^{choke} - P_A^{Mon}\right)}{2} Q_A^{Mon} = \frac{\left(1 - \frac{1}{2}\right)}{2} \frac{1}{2} = \frac{1}{8}$$

While the producer surplus is given by (area B):

$$PS_{Mon} = P_A^{Mon} Q_A^{Mon} = \left(\frac{1}{2}\right) \frac{1}{2} = \frac{1}{4}$$

Total economic surplus is then (area A + B)

$$TS_{Mon} = CS_{Mon} + PS_{Mon} = \frac{1}{8} + \frac{1}{4} = \frac{3}{8}$$

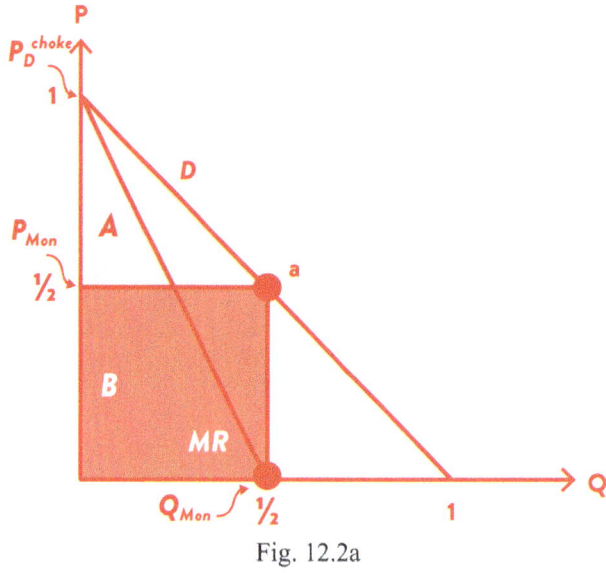

Fig. 12.2a

b. Then a foreign producer enters the market, creating a duopoly. The revenue (which, with zero marginal cost, is the same as profits) for the two firms becomes:

$$\pi_A = (1 - Q_A - Q_B)Q_A$$

$$\pi_B = (1 - Q_A - Q_B)Q_B$$

Maximizing profits for firm A with respect to its quantity produced we find that:

$$\frac{\partial \pi_A}{\partial Q_A} = 1 - 2Q_A - Q_B = 0$$

Firm A's reaction function then becomes:

$$Q_A = \frac{1 - Q_B}{2} = R_A(Q_B)$$

Due to symmetry, the reaction function of B must then be:

$$Q_B = \frac{1 - Q_A}{2} = R_B(Q_A)$$

Equilibrium occurs where the reaction functions intersect, which gives:

$$Q_A^{Duo} = Q_B^{Duo} = \frac{1}{3}, Q^{Duo} = \frac{2}{3}$$

With $P = 1 - Q$ and $Q = Q_A^{Duo} + Q_B^{Duo}$ the duopoly price is:

$$P_{Duo} = \frac{1}{3}$$

The Cournot duopoly is given by point b i Fig. 12.2b. The consumer surplus is now (area A):

$$CS_{Duo} = \frac{\left(P_D^{choke} - P^{Duo}\right)}{2} Q^{Duo} = \frac{\left(1 - \frac{1}{3}\right)}{2} \frac{2}{3} = \frac{2}{9}$$

The producer surplus of the domestic firm A is (area B):

$$PS_A^{Duo} = P^{Duo} Q_A^{Duo} = \left(\frac{1}{3}\right)\frac{1}{3} = \frac{1}{9}$$

The producer surplus of the foreign firm B is area C. Total economic surplus in the home country (area A + B) is:

$$TS_{Duo} = CS_{Duo} + PS_A^{Duo} = \frac{2}{9} + \frac{1}{9} = \frac{1}{3}$$

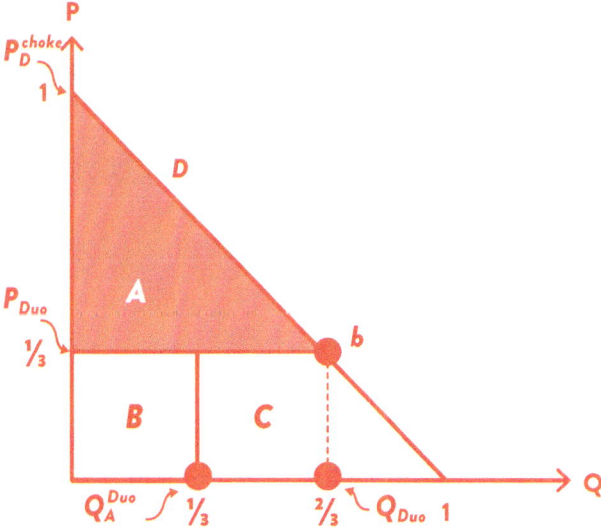

Fig. 12.2b Cournot

If we compare this duopoly solution with the monopoly outcome found in part (a), we see that the economic surplus has decreased because of the foreign entry:

$$TS_{Mon} = \frac{3}{8} > TS_{Duo} = \frac{1}{3}$$

This happens because the foreign producer has displaced some of the profits that previously went to the home firm:

$$PS_{Mon} = \frac{1}{4} > PS_A^{Duo} = \frac{1}{9}$$

This effect outweighs the gain to consumers, who experience higher quantity and lower price, and therefore higher consumer surplus:

$$CS_{Duo} = \frac{2}{9} > CS_{Mon} = \frac{1}{8}$$

c. What if A is the Stackelberg leader? We substitute Q_B with B's reaction function, $R_B(Q_A) = 0.5(1 - Q_A)$, in A's profit function and find that:

$$\pi_A = \left(1 - Q_A - \frac{1 - Q_A}{2}\right)Q_A = 0.5Q_A - 0.5Q_A^2$$

Maximising with respect to Q_A gives us:

$$\frac{\partial \pi_A}{\partial Q_A} = 0.5 - Q_A = 0$$

This gives the leader's quantity as:

$$Q_A^{Lead} = \frac{1}{2}$$

Plugging this into the follower's reaction function, $R_B(Q_A) = 0.5(1 - Q_A)$, we find the follower's quantity as:

$$Q_B^{Follow} = \frac{1}{4}$$

Total quantity and price are then:

$$Q^{Sta} = Q_A^{Lead} + Q_B^{Follow} = \frac{3}{4}, P^{Sta} = \frac{1}{4}$$

The Stackelberg equilibrium is illustrated as point c Fig. 12.2c. Consumer surplus (area A) is now:

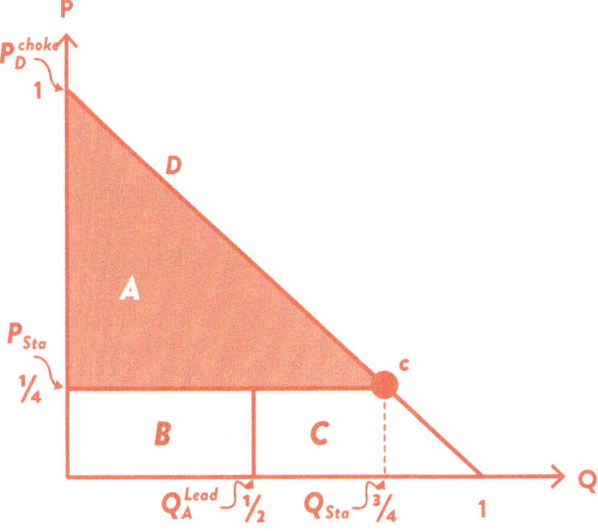

Fig. 12.2c Stackelberg

$$CS_{Sta} = \frac{\left(P_D^{choke} - P^{Sta}\right)}{2} Q^{Sta} = \frac{\left(1 - \frac{1}{4}\right)}{2} \frac{3}{4} = \frac{9}{32}$$

The producer surplus of the home firm, firm A, is (area B):

$$PS_A^{Sta} = P^{Sta} Q_A^{Sta} = \left(\frac{1}{4}\right)\frac{1}{2} = \frac{1}{8}$$

The producer surplus of the foreign firm, the follower B, is area C. Total economic surplus in the home country is now (area A + B):

$$TS_{Sta} = CS_{Sta} + PS_A^{Sta} = \frac{9}{32} + \frac{1}{8} = \frac{13}{32}$$

We see that $TS_{Sta} > TS_{Mon} > TS_{Duo}$ (since $13/32 > 3/8 > 1/3$).

The increase in social welfare with the home producer as Stackelberg leader, compared to the Cournot duopoly, is due to two factors.

First, the total quantity produced is higher and the market price is lower. This benefits consumers and improves sthe economic surplus because the outcome moves closer to the perfectly competitive solution, which is socially optimal.

Second, the home producer, as the Stackelberg leader, captures a larger share of the market, so less producer surplus flows out of the country.

$$CS_{Sta} = \frac{9}{32} > CS_{Duo} = \frac{2}{9}$$

$$PS_A^{Sta} = \frac{1}{8} > PS_A^{Duo} = \frac{1}{9}$$

12.3 Lower Costs, Higher Food Prices?
a. From Math Box 12.1, we see that in the Cournot equilibrium, firm A produces:

$$Q_A^{Duo} = \frac{1}{3}(1 - 2c_A + c_B) \quad \text{A's production in Cournot equilibrium}$$

And due to symmetry, firm B's production must be:

$$Q_B^{Duo} = \frac{1}{3}(1 - 2c_B + c_A) \quad \text{B's production in Cournot equilibrium}$$

Total production in Cournot equilibrium is therefore:

$$Q_{Duo} = Q_A^{Duo} + Q_B^{Duo} = \frac{1}{3}(2 - c_A - c_B)$$

Which plugged into the demand function gives the price:

$$P_{Duo} = 1 - Q_{Duo} = \frac{1}{3}(1 + c_A + c_B)$$

So far, this is just a repetition of what is stated in Math Box 12.1. But now let's insert the cost information from the exercise. If the two firms initially have the same purchase price, $c_A = c_B = 0.25$, then we get:

$$Q_A^{Duo} = Q_B^{Duo} = \frac{1}{3}(1 - 2(0.25) + 0.25) = 0.25$$

Total production is then:

$$Q_{Duo} = Q_A^{Duo} + Q_B^{Duo} = 0.5$$

And the price:

$$P_{Duo} = 1 - Q_{Duo} = 0.5$$

Figure shows the Nash equilibrium, where the two reaction functions intersect at point a, giving $Q_A = Q_B = 0.25$, and therefore a total production of $Q = 0.5$. Figure shows the market price and quantity for this equilibrium.

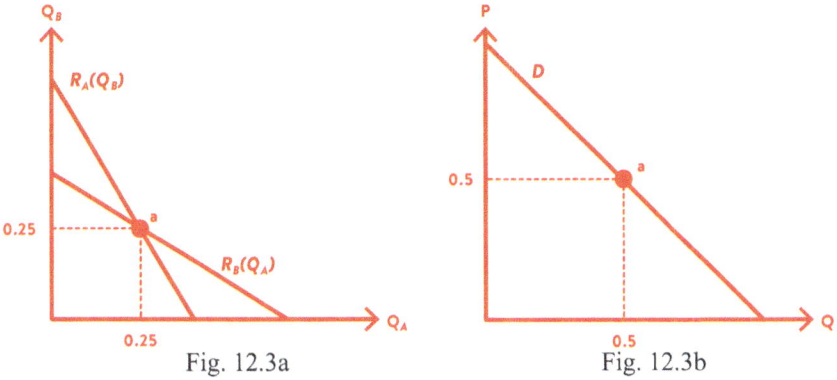

Fig. 12.3a   Fig. 12.3b

b. If firm A gets a low purchase price, $c_A^{low} = 0.2$, and B the worse deal, $c_B^{high} = 0.6$, the quantity produced by firm A is:

$$Q_A^{Duo} = \frac{1}{3}(1 - 2(0.2) + 0.6) = 0.4$$

While the equilibrium production of B is:

$$Q_B^{Duo} = \frac{1}{3}(1 - 2(0.6) + 0.2) = 0$$

Since only firm A will produce now, the price is:

$$P_{Duo} = 1 - Q_A^{Duo} = 1 - 0.4 = 0.6$$

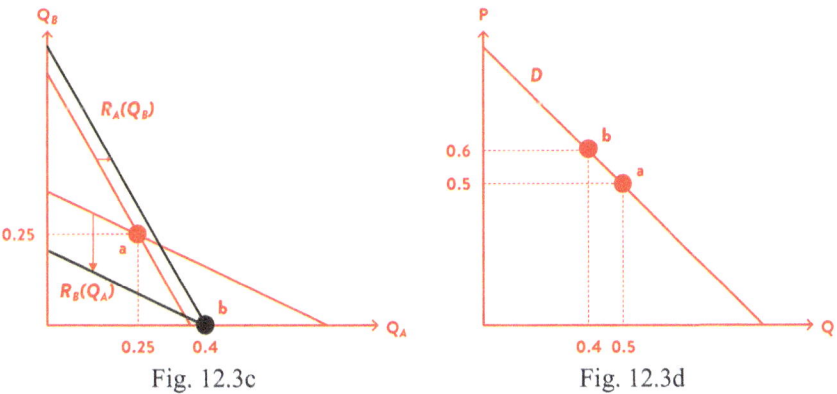

Fig. 12.3c   Fig. 12.3d

Figure 12.3c shows how the equilibrium changes because of the changes in purchase prices. The reaction function for A shifts outward, while that of B shifts inward, moving the equilibrium from point *a* to point *b*. Here, we see that $Q_A = 0.4, Q_B = 0$. Firm A effectively becomes a monopolist, and the total production in the market is now $Q = 0.4$. Figure 12.3d shows the market price and quantity for this equilibrium, where we observe that the quantity has decreased and, consequently, the price has increased due to the discriminatory pricing between the two producers.

12.4 Advertising, Customer Loyalty and Profits

a. From Math Box 12.3 on Bertrand competition, we know that when the firms have symmetric demand and the marginal cost is zero $(c = 0)$, their reaction functions become:

$$p_A = \frac{1 + p_B}{4} = R_A(p_B) \quad \text{Reaction function of Alpha before advertisement}$$

$$p_B = \frac{1 + p_A}{4} = R_B(p_A) \quad \text{Reaction function of Beta before advertisement}$$

In Bertrand equilibrium, the prices are therefore:

$$P_A^{Duo} = P_B^{Duo} = \frac{1}{3}$$

This is shown in Fig. 12.4, where the equilibrium is point a. Since the profits (which here equals revenue) of Alpha are given by $\pi_A = p_A(1 - 2p_A + p_B)$, then with $p_A = p_B = 1/3$ we have profits (symmetric for Beta):

$$\pi_A = \pi_B = \frac{1}{3}\left(1 - 2\left(\frac{1}{3}\right) + \frac{1}{3}\right) = \frac{2}{9} \approx 0.22 \quad \text{Profits before advertisement}$$

b. Alpha's advertising campaign makes their customers more loyal, which we interpret as them being more willing to continue buying Alpha's product even if p_A increases. With the demand function $Q'_A = 1 - 1.5p_A + p_B$ we can find Alpha's reaction function by maximising the profit $\pi'_A = p_A(1 - 1.5p_A + p_B)$:

$$\frac{\partial \pi'_A}{\partial p_A} = 1 - 3p_A + p_B = 0$$

This gives the reaction function:

$$p_A = \frac{1 + p_B}{3} = R'_A(p_B) \quad \text{Reaction function of Alpha after advertisement}$$

In Fig. 12.4, we see that Alpha's reaction function has shifted outward (it now crosses the horizontal axis at $p_B = 0 \Rightarrow p_A = 1/3$). Beta's reaction function remains as before, described in part (a). We find the equilibrium price by combining the two reaction functions, here substituting $R_B(p_A)$ for p_B:

$$p_A = \frac{1 + \frac{1+p_A}{4}}{3}$$

$$3p_A - \frac{1}{4}p_A = 1 + \frac{1}{4}$$

$$\frac{11}{4}p_A = \frac{5}{4}$$

$$P_A^{Duo} = \frac{5}{11} \approx 0.45$$

Plugging this into Beta's reaction function we get:

$$P_B^{Duo} = \frac{1 + \frac{5}{11}}{4} = \frac{4}{11} \approx 0.36$$

This is illustrated as point b in Fig. 12.4. We see that Alpha's advertising causes Beta to also choose to increase its price.

What are the implications for the firms' profits? We substitute the equilibrium prices found above into the profits expressions:

$$\pi_A' = \frac{5}{11}\left(1 - 1.5\left(\frac{5}{11}\right) + \frac{4}{11}\right) = \frac{75}{242} \approx 0.31 \quad \text{Alpha's profits after advertisement}$$

$$\pi_B' = \frac{4}{11}\left(1 - 2\left(\frac{4}{11}\right) + \frac{5}{11}\right) = \frac{32}{121} \approx 0.26$$

We observe that the advertisement has increased the profits of both firms: $\pi_A' > \pi_A$ and $\pi_B' > \pi_B$.

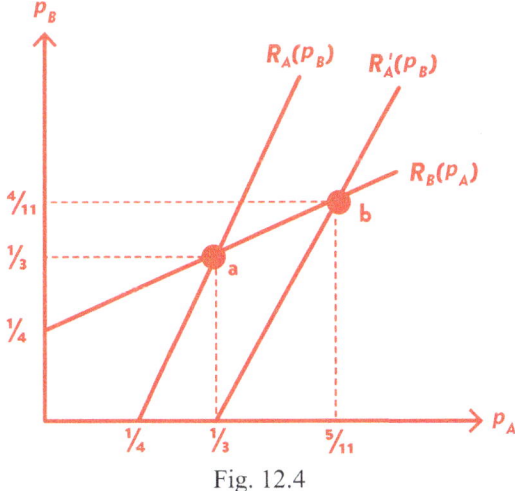

Fig. 12.4

The advertising thus enables Alpha to raise its price. But the analysis also shows that Alpha's advertising causes Beta to increase its price, allowing Beta to achieve higher profits as well. The reason why the price of good B also goes up is that prices are strategic complements in Bertrand competition: if one firm raises its price, the other will follow by raising its price too! The theory offers some interesting insights, doesn't it?

12.5 Bertrand Leader

a. From Math Box 12.3, we know that Beta's reaction function in this case (with $c_B = 0$) is:

$$p_B = \frac{1 + p_A}{4} = R_B(p_A)$$

We can then substitute this into Alpha's profit function:

$$\pi_A = p_A(1 - 2p_A + p_B)$$

Maximizing with respect to p_A:

$$\frac{\partial \pi_A}{\partial p_A} = 1 - 4p_A + \frac{1}{4} + \frac{1}{2}p_A = 0$$

$$\frac{7}{2}p_A = \frac{5}{4}$$

$$p_A^{Lead} = \frac{10}{28}$$

From Math Box 12.3, we know that with $c = 0$, when the two firms choose prices simultaneously, the equilibrium prices are $p_A^{Duo} = p_B^{Duo} = 1/3$. Since $p_A^{Lead} = 10/28 > 1/3$, we can conclude that the leader chooses to increase its price compared to the situation with simultaneous moves.

b. In Fig. 12.5, the equilibrium in the leader–follower model is given by point b. We can find the price of good B by substituting the leader's price p_A^{Lead} into Beta's reaction function:

$$p_B^{Follow} = \frac{1 + p_A^{Lead}}{4} = \frac{1 + \frac{10}{28}}{4} = \frac{19}{56}$$

Intuitively, the leader chooses to raise its price, and the follower follows by also raising its price. The movement from point a to b along $R_B(p_A)$ means higher profits for Beta. So, in this model, sequential moves result in higher profits for both producers, in contrast to the quantity competition situation (the Stackelberg model), where the leader benefits at the expense of the follower.

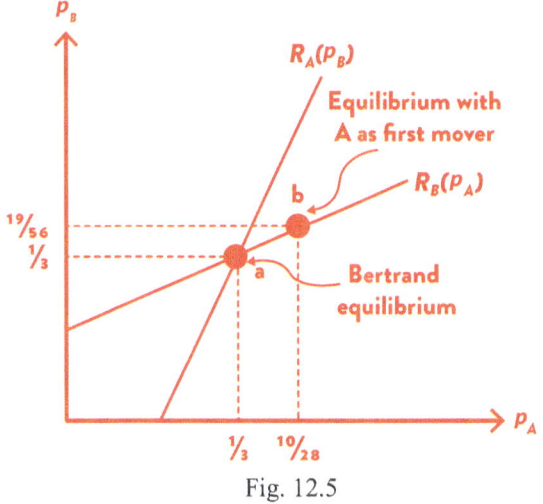

Fig. 12.5

Chapter 13 Game Theory

13.1 Two Games, abcd

a. A Prisoner's Dilemma is a situation where the players fail to achieve what is jointly best for them. Table 13.2 in the textbook shows this with a numerical example. Here, you should think more generally about the relationship between the different payoffs needed for the game to be a Prisoner's Dilemma.

We are told that high price "d, d" yields the highest total payoff. But in a Prisoner's Dilemma, it is tempting to deviate from this strategy. That is, for both players, b > d: given that the other player chooses the high price, it pays off for you to choose the low price.

In other words: b > d.

At the same time, in a Prisoner's Dilemma, if one player has deviated from the strategy that yields the highest total payoff, the other player also wants to deviate. That is, a > c: your best response to the other player choosing a low price is for you to do the same.

In other words: a > c.

Thus, the ranking is:

$$b > d > a > c$$

b. The Stag Hunt game is characterized by having two equilibria, where both players on which equilibrium is best. From exercise 13.1, we are given that "Green, Green" yields the highest total payoff.

What is required to make "Green, Green" an equilibrium? If one player chooses Green, the other must also choose Green, which means that d > b. At the same time, since the game should have two equilibria, the best response to Brown must be Brown, meaning a > c.

In other words:

a > c and d > b.

Table 13.1a Table 13.1a Prisoner's Dilemma, abcd

BETA BOOKS

		Low price	High price
ALPHA BOOKS	**Low price**	√a, a√	√b, c
	High price	c, b√	d, d

Table 13.1b Table 13.1b Stag Hunt, abcd

EAST

		Brown	Green
WEST	**Brown**	√a, a√	b, c
	Green	c, b	√d, d√

c. In Table 13.1c below, we swap the values of a and b compared to how they are in the main textbook (where a = 1, b = 2), setting a = 2 and b = 1. Does this change the game? No, they can be swapped without affecting the game: we still have two equilibria, "Brown, Brown" and "Green, Green".

Table 13.1c Table 13.1c Stag Hunt

EAST

		Brown	Green
WEST	**Brown**	√2, 2√	1, 0
	Green	0, 1	√3, 3√

In the current context, with the choice between brown and green technology, it seems reasonable to have b > a. Look at the table above and assume both produce with brown technology. The numbers imply that if one player switches from brown to green technology, both parties would lose, which sounds a bit strange! Then the numbers in the textbook make more sense, at least in an environmental context: For example, if East makes a green shift and contributes to a better environment, West will benefit from it—they get an environmental gain for free!

13.2 Back in the Office After the Pandemic

a. The situation after the pandemic, as described in Table 13.2a in the exercise, is a Prisoner's Dilemma: While both workers would benefit from going to the office, they end up staying at home. The reason is that it's tempting to leave the office duties to your colleague, so there's a free-rider problem at work. We can see this from the fact that if, say, Beth goes to the office, Audrey derives a utility of 2 from also going there, but a higher utility of 3 of staying at home.

Table 13.2a   Table 13.2a Prisoner's Dilemma

BETH

		Home	Office
AUDREY	**Home**	$\sqrt{1}$, $1\sqrt{}$	$\sqrt{3}$, 0
	Office	0, $3\sqrt{}$	2, 2

One way to interpret this is that certain office duties can be left to the other person. Audrey, for example, can enjoy the benefits of working from home—such as avoiding the commute—while relying on Beth to handle in-office tasks. The same reasoning applies to Beth. As a result, the equilibrium outcome is that both choose to work from home.

b. The game with the free lunch is shown in Table 13.2b. It is now more attractive to show up to work, as evidenced by the higher utility Audrey and Beth experience when they're both at the office (3,3), working and having lunch together. Of course, lunch at work does not affect utility if you're not there, so the utility of staying at home is unchanged (1,1).

Given that the other one is at the office, it is now more tempting to go to work than to stay at home, 3 > 2, and so one equilibrium is that both go to the office. However, both staying at home is also an equilibrium: Given that you think your colleague will stay at home, you would rather also stay at home rather than go to work (and eat the lunch alone). Offering lunch thus transforms the game into a Stag Hunt, with multiple equilibria. Note that while both going to the office is clearly the better equilibrium, and offering lunch may do the trick, there is no guarantee that this will actually be the outcome of the game: the firm may still end up in a

Table 13.2b Coordination game

		BETH	
		Home	Office
AUDREY	Home	√1, 1√	2, 0
	Office	0, 2	√3, 3√

situation where workers choose to work from home (in which case the manager might want to consider additional measures).

c. What if Audrey makes a first-mover choice? This transforms the game into sequential moves, and we can use the extensive form to show the equilibrium, see Fig. 13.2.

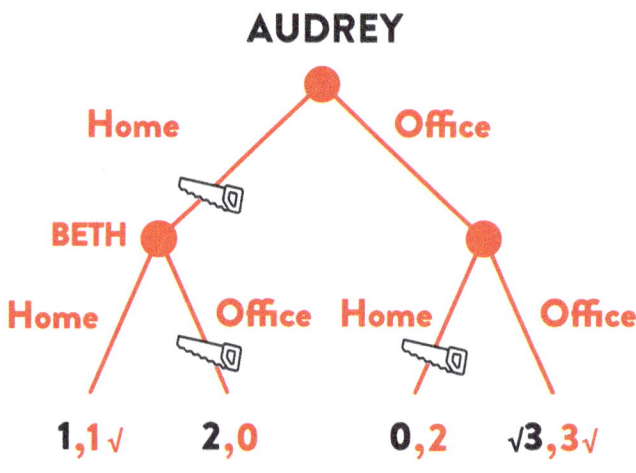

Fig. 13.2

Using backward induction, we see that if Audrey chooses to stay at home, Beth best choice is to also stay at home. We can use the saw on the "Home, Office" branch.

If, on the other hand, Audrey chooses to go to work, Beth will do the same. Accordingly, we can use the saw on the "Office, Home" branch.

Considering the remaining choices "Home, Home" and "Office, Office", Audrey clearly prefers the latter, so she uses the saw on the former one, and we end up with an equilibrium where both go to the office. Giving Audrey a leadership role has therefore solved the problem of getting the workers back to work!

13.3 Anna and Brian on Their First Date

We insert the given utility levels into a matrix model, as shown in Table 13.3.

Table 13.3 Table 13.3 Anna and Brian on their first date

		BRIAN Cinema	Café
	Cinema	√1, 2√	0, 0
ANNA			
	Café	0, 0	√2, 1√

Anna thinks that if Brian chooses café, she will do the same, and if Brian chooses cinema, she will also choose the same. Brian thinks the same way. As we can see, this game has two Nash equilibria: "Cinema, Cinema" and "Café, Café".

This is a classic game, often called Battle of the Sexes since even in the original formulation from the 1950s, it was presented as a conflict of interest between a couple.

What is clear from this game is that Anna and Brian will spend the evening together, but not where.

This game resembles the Chicken Game described in chapter 13.4 of the textbook. But unlike the Chicken Game, where the point is to avoid doing the same thing, the point in the Battle of the Sexes is to avoid doing different things.

13.4 More Diet Books?

a. When we allow publishers to release two books each year, where the profits from releasing two books is the sum of releasing in January and in May, we get a payoff matrix as shown in Table 13.4.

Table 13.4 More diet books

		BETA BOOKS January	May	Jan. & May
	January	1.5, 1.5	3, 2	1.5, 3.5√
ALPHA BOOKS	May	2, 3	1, 1	1, 4√
	Jan. & May	√3.5, 1.5	√4, 1	√2.5, 2.5√

To understand the matrix, look at the situation where Beta Books releases a book only in January, that is the first column. If Alpha Books releases its books in both January and May, we are in the bottom-left cell. Here, Beta Books earns a profit of 1.5 from its January sales while Alpha Books earns a profit of 1.5 from January plus 2 from May, totalling 3.5.

Similarly, consider the case where Beta Books only releases a book in May, that is, the second column. If Alpha Books now releases the books in both months, it earns a profit of 3 from January and 1 from May, a total of 4. In contrast, Beta Books only gets a payoff of 1 in this case.

Finally, if Beta Books releases books in both months, we're in the third column. If Alpha Books also publishes the books in both months, we're in the bottom right cell, and both firms earns 2.5, where 1.5 comes from the January sales and 1 from the May sales.

Using the tick-off method, we see that the Nash equilibrium in this case is to release two books per year: this is the dominant strategy for both publishers.

b. Since there is one dominant strategy for both publishers in this game, and thus one Nash equilibrium—namely releasing two diet books per year—this will also be the equilibrium in a sequential game (regardless of who moves first). Therefore, there is no advantage to moving first, and the publishers will not want to spend money signalling future releases to their competitor (for example, by buying advertising space in magazines).

13.5 Brian and His Mother Again
a. We set up a game tree with Brian as the first mover, as shown in Fig. 13.5a. Brian's choice is whether to work or not, while Audrey's choice is whether or not to help her son.

We use backward induction and first look at what Audrey will do if Brian chooses to work. Audrey's best response in this case, marked by the second number at the bottom, is to help, which gives her a utility of 3, while not helping yields only utility 2. We can therefore cut off the branch "Help? No" as a response to Brian's "Work? Yes".

What if Brian chooses not to work? We again see that Audrey's best response is to help, which gives her utility 0.25 compared to zero if she does not support him. We can therefore cut off the branch "Help? No" as a response to Brian's "Work? No". Audrey's best response is thus always to help her son. The threat of not supporting him if he does not find a job is therefore not credible.

Brian's choice is simple: He compares "Work? No, Help? Yes" with "Work? Yes, Help? Yes" and chooses not to work, which gives him utility 3 (while working would give him utility 2).

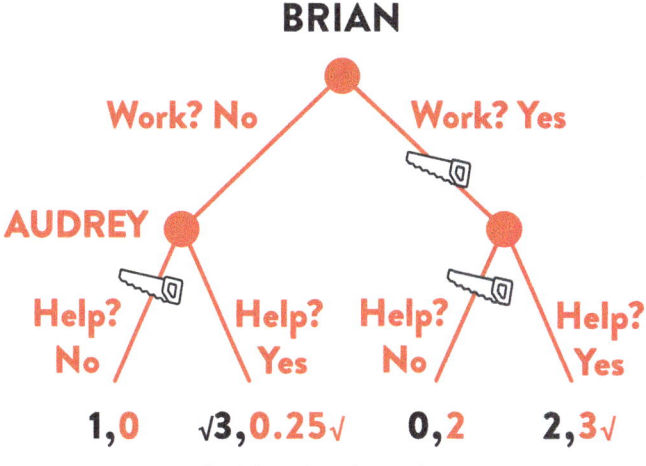

Fig. 13.5a Starting point

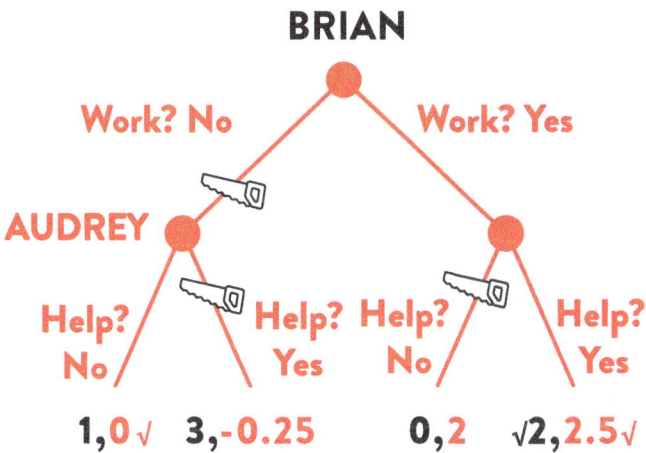

Fig. 13.5b When giving is more expensive

b. Now supporting her son costs an extra 0.5, and Audrey's utility from helping goes down, as shown in Fig. 13.5b. In this case, helping her son gives her a utility of minus 0.25 if he does not work, and 2.5 if he works.

Note that Audrey's best response to Brian choosing not to work is now not to help, since helping him results in a utility of −0.25, while not helping yields utility zero. We can therefore use the saw on the branch "Help? Yes" when Brian chooses "Work? No".

Offering financial support is also more costly if Brian works, but here she has more leeway, and will still choose to help Brian if he chooses to work. As before, we cut off the branch "Help? No" if Brian has chosen "Work? Yes".

The threat of not helping Brian if he does not work is now credible.

Brian now faces the choice between not working, which will lead his mother to choose not to help, and working, which will lead her to help him. The choice is clear: working gives him utility 2, while not working gives utility 1, so Brian chooses to work.

It is interesting to note that by putting herself in a position where she must take on credit card debt—which does not give Audrey any positive utility by itself (her utility goes down when she chooses to help and is unchanged if she doesn't offer any help)—it can still be beneficial for her. The point is that by putting herself in this difficult situation, Audrey creates a credible threat not to offer help if Brian does not find a job. And with this credible threat, Brian chooses to work, which gives Audrey higher utility. Pretty cool, right?

The manufacturer's authorised representative in the EU is Springer
Nature Customer Service Centre GmbH, Europaplatz 3, 69115 Heidelberg,
Germany. If you have any concerns regarding our products, please
contact ProductSafety@springernature.com

Printed and bound by CPI Group (UK) Ltd, Croydon, CR0 4YY

04/05/2026

02102349-0003